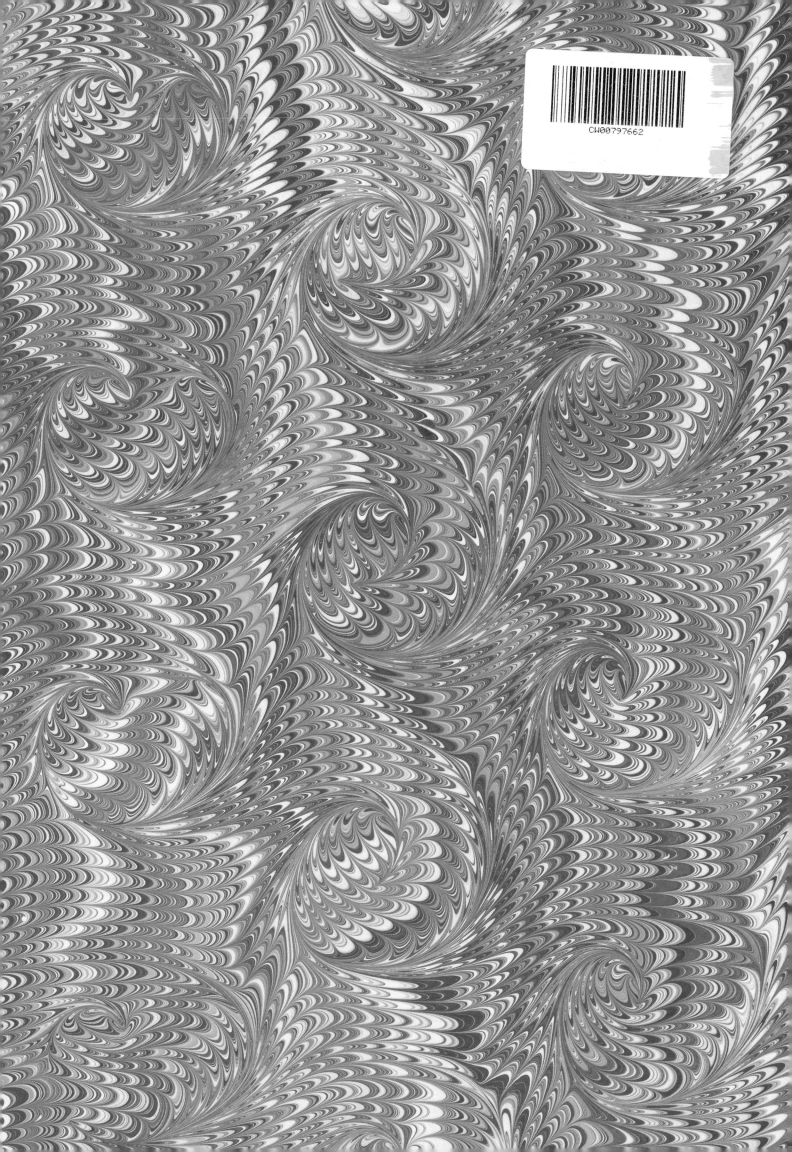

GERMANY'S TIGER TANKS
D.W. TO TIGER I

By Thomas L. Jentz
GERMANY'S TIGER TANKS: Vol.1 - D.W. to TIGER I
GERMANY'S TIGER TANKS: Vol.2 - V.K.45.02 to TIGER II
GERMANY'S TIGER TANKS: Vol.3 - TIGER I & II: COMBAT TACTICS
GERMANY'S PANTHER TANK: THE QUEST FOR COMBAT SUPREMACY
TANK COMBAT IN NORTH AFRICA
PANZERTRUPPEN: 1933-1942
PANZERTRUPPEN: 1943-1945

GERMANY'S TIGER TANKS

D.W. to TIGER I

Thomas L. Jentz & Hilary L. Doyle

Schiffer Military History
Atglen, PA

Book Design by Robert Biondi.

Copyright © 2000 by Thomas L. Jentz.
Library of Congress Catalog Number: 99-66300.

All rights reserved. No part of this work may be reproduced or used in any forms or by any means – graphic, electronic or mechanical, including photocopying or information storage and retrieval systems – without written permission from the copyright holder.

Printed in China.
ISBN: 0-7643-1038-0

We are interested in hearing from authors with book ideas on related topics.

Published by Schiffer Publishing Ltd.
4880 Lower Valley Road
Atglen, PA 19310 USA
Phone: (610) 593-1777
FAX: (610) 593-2002
E-mail: Schifferbk@aol.com.
Visit our web site at: www.schifferbooks.com
Please write for a free catalog.
This book may be purchased from the publisher.
Please include $3.95 postage.
Try your bookstore first.

In Europe, Schiffer books are distributed by:
Bushwood Books
6 Marksbury Ave.
Kew Gardens
Surrey TW9 4JF
England
Phone: 44 (0)208 392-8585
FAX: 44 (0)208 392-9876
E-mail: Bushwd@aol.com.
Free postage in the UK. Europe: air mail at cost.
Try your bookstore first.

Contents

Introduction 6

Chapter 1: DESIGN AND DEVELOPMENT 9
    1.1: Durchbruchswagen 10
    1.2: VK 30.01 12
    1.3: VK 36.01 17

Chapter 2: PANZERKAMPFWAGEN VI (PORSCHE) 23
    2.1: Porsche Typ 100 23
    2.2: Porsche Typ 101 25
    2.3: Porsche Typ 102 29

Chapter 3: PANZERKAMPFWAGEN TIGER AUSF.E 30
    3.1: Development 30
    3.2: Description 34
    3.3: Tiger I Production 67
    3.4: Modifications During Production 71
    3.5: Modifications After Issue 164

Appendices 169
    Appendix A1: Pz.Kpfw.VI (VK 4501 P) (Ausf.P) 171
    Appendix A2: Porsche "Typ 101A" 172
    Appendix A3: Porsche Design Data 173
    Appendix B1: Pz.Kpfw.VI (VK 4501 H) (Ausf.H1) (Tiger) 175
    Appendix B2: Panzerkampfwagen Tiger I Data 177
    Appendix B3: Tiger E Technical Specifications 179
    Appendix C: Part Numbers 183
    Appendix D: Armor Specifications 185
    Appendix E: Comparison Trials 189
    Appendix F: Camouflage Paint and Zimmerit 190

Introduction

As Hilary and I set out to create this series of books on Germany's Tiger tanks, we had several goals in mind. The first goal was to set the record straight by only using information extracted directly from original source material. Another goal was to create an unequaled reference work that would be useful for correctly interpreting photos found in both this and other publications. A third goal was to provide modelers with sufficient information to accurately model the top, bottom, and all sides of a Tiger, even if they only had a single photographic view of the Tiger they wanted to model. And last but not least, a primary goal was to present a history of how the Tigers fared in combat as related by those who fought in them.

Over 30 years of intensive research went into finding the original documents needed to create this history of the development, characteristics, and tactical capabilities of the Tiger. An exhaustive search was made for surviving records of the design/assembly firms (including Krupp, Henschel, Porsche, and Wegmann), the **Heereswaffenamt**, the **Generalinspekteur der Panzertruppen**, the D656 series of manuals on the Tiger, and the war diaries with their supporting reports from German army units. This was supplemented by our collecting hundreds of photos and climbing over, under, around, and through almost every surviving Tiger I.

Not one single bit of information has been derived from other published books. If it couldn't be found in an original document, photograph, or surviving Tiger, it isn't in this book. The sources of information are stated in the text. Those desiring footnotes and an extensive bibliography are advised to look elsewhere. The goal is to create an accurate record of events, not prove that a list of over 200 publications can be accumulated.

Events occur in only one way. A clear picture of these events cannot be obtained from stories, military intelligence reports, expert opinion, analyses of raw data, or secondary sources. Either the events were recorded by a participant or they weren't. Unfortunately some key records, such as Krupp's correspondence on the design of their turret for the period from June to December 1941, didn't survive the war. Thus the story of the evolution of the turret designed by Krupp for the Porsche Tiger will never be known. No amount of conjecture, supposition, reasoning, or opinion can fill gaps like this in the original records. Even though participants in the design of the Panzers made occasional mistakes in their records and sometimes slanted their reports to be more or less favorable, they were the only direct observers. Any gross errors that may have been made in an original report can be discovered by careful examination of the end product and comparison to other independently created original reports and drawings.

Many errors have crept into postwar publications through the use of popular nicknames for the German Panzers. If an author doesn't even know the correct name for a Panzer, what else is wrong with his information? The most commonly used names are often the most misleading. To set the record straight, the names found in the original documents have been used in the text and listed in the introductory sections of chapters. As can be seen from these lists, there was no single official name. New names simply evolved with time during the war. However, this does not give us license as postwar historians to fabricate, propagate, use, and spread misnomers. The original German names for the Panzers and component parts are shown throughout the text in bold print. When appropriate, the equivalent wartime American military terms are added in parentheses.

Introduction

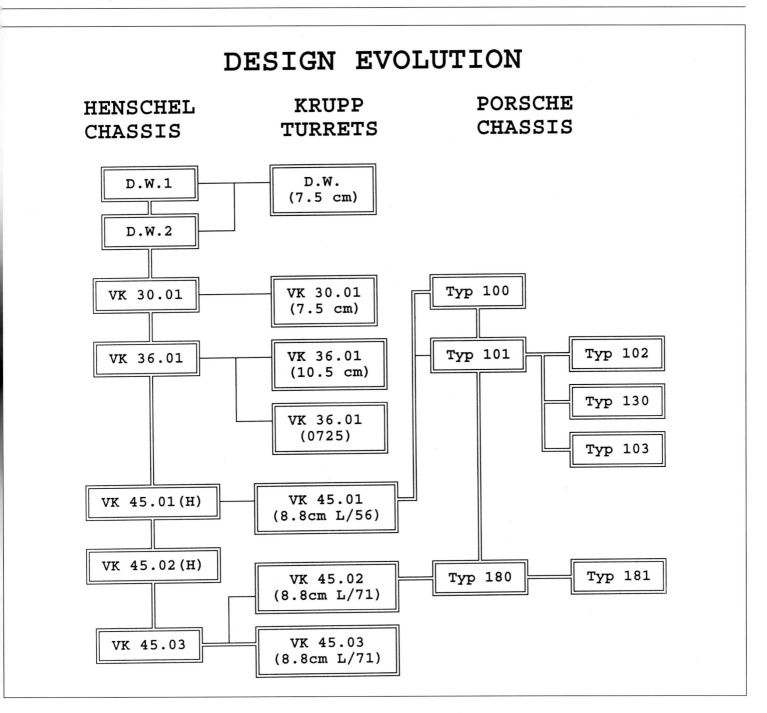

Gruppen-Nummern were used during the war to identify drawings of complete vehicles, subcomponents, and parts belonging to each Tiger. As parts were made, they were labeled by either stamping the drawing number into the surface or casting the number onto castings. The numbers were listed in manuals to identify specific replacement parts needed to repair each different **Ausfuehrung** (model). The **Gruppen-Nummern** assigned by **Wa Pruef 6** (or Porsche) to the Tiger series and its predecessors were:

Model	Chassis	Turret
D.W.1 u. 2	33600	15230
VK 30.01	39300	39350
Pz.Sfl.V	39500	N/A
Typ 100	100	860
VK 36.01	4200	???
VK 45.01(P)	101	860
VK 45.01(H)	2700	2760
VK 45.02(P)	180	48010
VK 45.03	49500	50600

While only a limited application still remains for using these part numbers as an aid in restoring the few surviving Tigers, these numbers are the key to creating accurate models. The detailed **Gruppen-Nummern** lists for each Tiger reveal exactly which component part was adopted from a previous model and which parts were designed specifically for a new model (refer to Appendix C). Because it is very rare to have photographs of all views of a Tiger, and the surviving Tigers all have missing parts or postwar changes, these **Gruppen-Nummern** used in combination with the descriptions and drawings of the modifications are essential tools for creating accurate models.

This first volume of this three-volume set covers the history, development and production history of the Tiger tank variants from the idea's conception to the end of Tiger I production. This includes details on the development series known as the **D.W.**, **VK 30.01(H)**, **VK 30.01(P)**, **VK 36.01(H)**, **VK 45.01(P)** as well as the Tiger I. All of this illustrated with scale drawings by Hilary L. Doyle, combined with drawings, sketches, and photographs depicting external modifications as well as internal views.

A second volume, with Chapters 4 to 6 entitled <u>VK 45.02 to Tiger II</u>, continues coverage of the Tiger chassis developed by both Porsche and Henschel with Krupp turrets mounting the **8.8 cm Kw.K.43 L/71** gun. The third volume, with Chapters 7 to 9 entitled <u>Tiger I & II: Combat Tactics</u>, covers the operational characteristics, organization, units, issue records, tactical manuals, operational strength reports, and original experience reports.

The author would like to thank Dr.-Ing. Fetzer, Herr Loos, and Herr Meyer in the Bunderarchiv/Militaerarchiv in Freiburg for their many years of support in providing paths through the massive records. Also Frau Kuhl at the Bildarchiv in Koblenz for her assistance in obtaining photographs of exceptional quality. I was able to gain access to the inside of many of the surviving Tigers with the support and cooperation of the late Colonel Aubry and the current Colonel Olmer at the Musee Blindes, Saumur, Oberstleutnant a.D. Grundies at the Panzer Museum, Munster, Hauptmann Kosinski at the Wehrtechnischen Studiensammlung, Koblenz; and David Fletcher at The Tank Museum, Bovington. This allowed exact identification of each Tiger by its **Fahrgestell Nummer** (chassis number), a significant contribution toward determining the correct sequence of production modifications.

Hilary Louis Doyle was responsible for our efforts at tank diving at the museums. Since 1973, Hilary and I have been partners in digging out details on the design of the German tanks. Hilary is unequaled in the accuracy, detail, and skill applied to the scale drawings illustrating this and many other armor books. Thanks go to Stanley Thawley for proofreading the manuscript. Many other friends including Bill Auerbach, Werner Regenberg, Karlheinz Münch, Marcus Jaugitz, Tom Anderson, Georg Fancsovits, and Wolfgang Schneider have been very helpful in providing photographs.

Once again, special thanks go to my friend and mentor Walter Spielberger, the leader in researching and publishing books on the design of German tanks and other vehicles. In the area of technical descriptions, Walter has led the way. I have merely followed by adding and cleaning up a few details.

Tom Jentz
Germantown, Maryland
20 May 1999

1
Design and Development

The first mention of a Panzer in the 30 ton class was found in a report dated 30 October 1935 on **Offensive Abwehr von Panzerwagen** (offensive defense of tanks) by General Liese, head of the **Heeres Waffenamt**:

*The initial velocity of the 7.5 cm gun must be increased to about 650 meters/second to be effective against the Char 2 C, 3 C, and D. This type of increase requires the design of a completely new Panzer. Based on rough calculations, armor protection up to 20 mm thick (still not fully protected against 2 cm guns) would result in a weight of at least 30 tons. The **Oberbefehlshaber des Heeres** himself recently spoke against this type of tank.*

As a follow-up action, confirm that development of a medium Panzer weighing about 30 tons with a 7.5 cm gun with increased capability (muzzle velocity of 650 meters/second) can be dropped.

However, a high performance engine with adequate horsepower is needed before a heavy tank can be created. Dr. Maybach had met on 26 October 1935 with representatives from Wa Prw 6 (Oberstleutnant Phillips, Hauptmann von Wilcke, Dipl.Ing. Kniepkamp, Dipl.Ing. Blasberg, Dipl.Ing. Augustin) to discuss the development of a 600/700 horsepower motor:

In a lengthy discussion it was determined that a 700 horsepower gasoline engine with sustained output of about 650 horsepower should be developed. A 16-cylinder motor as proposed by Maybach couldn't be considered because value was placed on achieving the shortest possible installation length. Dr. Maybach emphasized that a step from 300 to 700 horsepower is still unconventially large and that it could mean a long development period. It needs to be very exactly tested if the maximum power for a 12-cylinder engine can be achieved between 600 and a peak of 700 horsepower. The entire position would be significantly simpler and surer if for the present one was satisfied with a peak performance of 600 horsepower and sustained output of 550 horsepower. Herr Blasberg then stated that based on this he was in favor of starting with a peak performance of 600 horsepower. Herr Augustin added that maybe later, 700 horsepower could still be required.

On 28 December 1935, development of a 600 horsepower engine was again discussed in a meeting between Herr Zabel of Maybach and representatives from Wa Prw 6:

Dipl.Ing. Augustin turned the discussion to the development of a 600 horsepower engine for heavy Panzers and noted that his opinion was that 600 horsepower will not be sufficient and that indeed it would be more correct to immediately develop a motor capable of 700 horsepower. Herr Zabel referred to the last discussion in which Dr. Maybach pointed out that 600 horsepower was the limit for a 12-cylinder engine which retained the previous design parameters, shape of the compression chamber, installation of the valves, etc.

When one considers that if a 700 horsepower engine would be required, one must probably go over to a 16 cylinder engine that naturally requires a larger installation length – estimated to be at least 50 cm longer. The increased performance in comparison to a 600 horsepower engine will then probably be wasted on the higher weight of the armor because the vehicle would naturally be about 1/2 meter longer, so that no advantage would be achieved with a 700 horsepower engine.

Oberstlt. Philipps noted that the Panzer, for which the 600 horsepower engine was intended, weighed about 30 tons, so that a power ratio of about 20 horsepower per ton would be available; which was completely favorable. Philipps believed that a higher weight would hardly be allowable when considering the Pionier bridging equipment. He held it to be important that the question of the maximum weight for Panzers be absolutely determined. Additionally, the development of a 600 horsepower engine is not so pressing because a contract for the design of a 30 ton Panzer has still not been awarded. Also the final decision still hasn't been made whether a 30 ton Panzer will be built at all. However, this should not mean that the development of the 600 horsepower engine should be delayed; to the contrary, in every case development should be completed.

Going for a 600 horsepower engine could be considered to be unbounded optimism, since at this time Maybach had yet to prove by bench test that 300 horsepower could be achieved with their 12-cylinder engines.

Wa Prw 6 had made the correct decision when they decided to abandon the use of commercially available high-torque, low rpm (1400-1600) aircraft engines for installation in the Panzers. Using high-torque aircraft engines meant that the other drive train components, from clutches through final drives, had to be designed to withstand this high torque and therefore were relatively heavy. Weight restrictions imposed by bridge load-bearing limits meant that additional weight in drive trains had to be paid for with smaller tanks, thinner armor, smaller armament, less ammunition, and/or fewer crew members. Therefore the firm of Maybach was given contracts to create compact, efficient, high-performance engines specifically designed for tanks. New high-performance engines incorporating features at the leading edge of available technology which didn't have the bugs worked out were continuously introduced. This led to frequent breakdowns until modifications to solve the reliability problems could be introduced during production runs. But keeping high-torque aircraft engines would have resulted in low lifespan and frequent breakdown of other drive train components, as experienced in British tanks.

Maybach Engines Designed for Heavy Tanks

Year	Name	Bore	Stroke	Cyl	HP	rpm	HP/L
1936	HL 120	105	115	12	300	3000	25.0
1937	HL 320	145	160	12	600		18.8
1938	HL 190				375		19.7
1938	HL 116	125	150	6	300	3000	25.9
1938	HL 224	125	145	12	600	3000	26.8
1939	HL 150	150	150	6	400	3400	26.7
1940	HL 174	125	130	12	500	3000	28.7
1941	HL 210	125	145	12	650	3000	31.0
1942	HL 230	130	145	12	700	3000	30.4

1.1 Durchbruchswagen

Over a year after initial deliberations on whether to build a 30 ton Panzer, in January 1937 Baurat Kniepkamp (an automotive specialist) of Wa Prw 6 (the tank design office of the ordnance department) asked Henschel to design a **Fahrgestell** (chassis). As with most Panzer designs, Wa Prw 6 laid down the specifications and selected the major components. It was then up to the contracted design shops, such as Abteilung Maschinenbau C of Henschel under Dr.-Obering. Aders, to create a working product within the restrictions of the specifications.

As in most of the previous Panzer series, Wa Prw 6 selected one firm to complete the detailed design for the chassis and a second firm to perform the detailed design work for the turret. Earlier, in November 1936, Wa Prw 6 had requested that Krupp create a conceptual design for a turret for a 30 ton Panzer mounting a 7.5 cm gun.

This Panzer was originally named **B.W. (verstaerkt)**. **B.W.**, the abbreviation for **Begleitwagen**, was the code name used for the **Pz.Kpfw.IV** series. It indicates that the **Pz.Kpfw.IV** was intended to be used to **Begleit** (escort) the lighter Panzers. The modifier **verstaerkt** (strengthened) implied a heavier armored Panzer having the same tactical application as the **Pz.Kpfw.IV**. On 12 March 1937, the name **B.W. (verstaerkt)** was officially changed to **I.W.**, the abbreviation for **Infanteriewagen**. However, this name didn't survive long, since on 28 April 1937, Wa Prw 6 directed: "The title for the **B.W. (verstaerkt)**, presently referred to as the **I.W.**, was now to be changed to **D.W. (Durchbruchs-wagen)**." The name implies that a new tactical role had been envisioned for these heavy Panzers to **durchbruch** (breach) the enemy defenses.

Much later, in November 1939, the original **D.W.** design was referred to as the **VK 30.01 alte Konstruktion** (old design). The older **D.W.** code name was retained and still used alongside the new **VK 30.01** designation.

1.1.1 Henschel Fahrgestell

As listed on their delivery plan dated 1 October 1937, Henschel Werkstatt "CS" planned to complete one **DW-Fahrzeug** with **Cletracgetriebe** (Cleveland track steering gear) and one **DW-Fahrzeug** with **Ueberlagerungsgetriebe** (multiple stage steering gear) in the second half of 1938. Henschel designated the former as **D.W.1** and the latter as **D.W.2**. Both the **D.W.1** and the **D.W.2 Erprobungs-Fahrgestelle** (experimental chassis) were completed and tested by Henschel.

1.1.1.1 D.W.1

The hull for the first **D.W.-Erprobungs-Fahrgestell**, designated **D.W.1** by Henschel, was fabricated from soft steel with plate thicknesses of 50 mm on the front, sides and rear, and 20 mm for the deck and belly. The hull was constructed in separate forward and rear sections that were bolted together with a vertical stiffener during assembly. The drive train consisted of a **Maybach HL 120 TR Motor**, a **Maybach Variorex Schaltgetriebe** (transmission) and **Cletrac** three stage steering gears capable of obtaining a maximum speed of 35 kilometers per hour. The torsion bar suspension supported six double, rubber-tired roadwheels per side on center-guide track with a 300 mm pitch.

Details on the design of the **D.W.1 Fahrgestell** were described by Dr. Aders in a report dated February 1945 as follows:

D.W.1 of 30 metric tons with maximum speed of 35 km/hr

Wanne (Hull) in two pieces. The rear piece was bolted on because steel mills couldn't roll the side walls in one long piece. Escape hatches in the bottom of the hull to the right front by the radio operator and to the left rear in the engine compartment. There was a door in the firewall which the crew could crawl through.

Armor skirts in front of the track drive wheel were proposed. These armor skirts could be raised and lowered by using a hand crank with reduction gears. An expedient model was shot at; the results were negative.

The wall thickness of the front, sides, and rear was 50 mm. Deck and belly were 20 mm thick.

Gleisketten (Tracks) with 300 mm pitch and needle-bearings (lubricated) were planned for backfitting rubber track pads.

Motor Maybach HL 120 with about 280 metric horsepower.

Kuehlung (Engine Cooling) Cooling air entered the top of side extensions (like the Pz.Kpfw.III). Radiators were mounted in the engine compartment. The fan drives were belt-driven, similar to the Z.W.38 (Pz.Kpfw.III Ausf.E).

Turm-Antrieb (Turret Drive) with worm gears driven directly from a piece mounted solidly on the main drive shaft crown. Insufficient lubrication was only later discovered on the VK 30.01.

Schaltgetriebe (Transmission) was a Maybach-Motorenwerk Variorex (semi-automatic).

Lenkgetriebe (Steering Gear) three Cletrac stages in series.

Chapter 1: Design and Development

Six brakes, four Ortlinghaus clutches. Friction value was estimated to be 0.2. Failure! Corrected and new design by Herr Kluge (and Herr Raht in Friedrichshafen with Maybach-Motorenwerk). Long development and trials in Sennelager. Hydraulic assisted operation from Tees. The cast iron housing broke twice, once because of foreign object caught in the gears. A cast steel housing was rapidly ordered.

Seitenwellen *(Side Drive Shafts)* easily steered with geared-clutches, various lengths.

Ketten-Antrieb *(Final Drives)* with 1 to 21.5 reduction, two hardened inner gears and one planetary gear system. Mounted on a shaft with a large flange on the hull side.

Bremsen *(Brakes)* outer backed brakes of our own design. The original Jurid linings were replaced by Goetze hard casting linings because of the unbearable smoke created. Operated by Teves-Lockheed linkage.

Tragrollen *(Return Rollers)* three rubber-tired wheels on shafts supported by flanges.

Laufrad-Kurbeln *(Roadwheel Support Arms)* individual forged pieces with shafts for the roadwheels fitted by shrinking. Novotext bushings in the hull side.

Stabfederung *(Torsion Bars)* entirely square cut and hollow (Roechling), set in behind each other. C-value of about 12-13 kg/mm. No spring breakage occurred.

Laufraeder *(Roadwheels)* cast steel discs with solid rubber double tires.

Stossdaempfer *(Shock Absorbers)* of special design from Boge u. Sohn. (Dr. Schroeder) mounted front and rear on the outside.

Anschlaege der Laufrad-Kurbeln *(Bumpstops for the Roadwheel Support Arms)* cast steel block with solid rubber cushions vulcanized to stick to steel. Four on the front and rear arms.

Leitrad-Achsen *(Idler Axle)* mounted in the hull, adjustable from inside without sheer pins. Turned inward.

Leitraeder *(Idler Wheels)* cast steel with solid rubber tires.

1.1.1.2 D.W.2

By mid-1937, Henschel had planned to construct a second **D.W.-Erprobungs-Fahrgestell** designated as **D.W.2**. Basically the **D.W.2** was similar to its predecessor the **D.W.1**, but with automotive improvements. Changes to this second **Fahrgestell**, **D.W.2,** included improvements in the steering gear, final drives, parking brakes, torsion bars, and tracks (pitch decreased to 260 mm, with matching changes for the drive sprockets).

Details on the design of the **D.W.2 Fahrgestell** were described by Dr. Aders in a report dated February 1945 as follows:

D.W.2 of 30 metric tons with maximum speed of 35 km/hr

Lenkgetriebe *(Steering Gear)* with magnetic clutches, three stage differential with reversed direction, therefore modification of the final drive, brakes, mounting for the gear box, and the torsion bars. Because the steering gear in the **Z.W.38** from Daimler-Benz had functioned successively only when driving on roads but not cross country, the large double radien was completely dropped and only the first Cletrac stage retained, operated by simple mechanical linkage.

Seiten-Antrieb *(Final Drives)* with a simple reduction gear before the planetary gear. Since gears weren't available with normal straight teeth, some with Lorenz-Sykes arrow teeth were used which were hardened like the inner gears for the **D.W.1**. Gear reduction was decreased to 1 to 12 and the direction reversed.

Haltebremse *(Stopping Brakes)* modified due to the reversed direction of the side drive shafts and lower torque.

Gleiskette *(Tracks)* with 260 mm pitch, otherwise the same as on the **D.W.1**. Significantly smoother ride.

Triebraeder *(Drive Wheels)* modified for 260 mm pitch tracks. Experimented with various mounting of the drive rollers, needle-bearings, and Novotext bushings.

Stabfederung *(Torsion Bars)* simple torsion bars abandoning soft springing. They turned out to be sufficient with a C value of 32 kg/mm.

1.1.2 Krupp Turm

In response to their request on 2 November 1936, Krupp sent Wa Prw 6 a conceptual design drawing for a turret for the **B.W. (verstaerkt)** on 22 February 1937. After reviewing the proposal, Dr. Olbrich, Wa Prw 6/IId, informed Krupp of the following decisions in a letter dated 12 March 1937:

*1. The **B.W.-Turm (verstaerkt)** receives the name: "I.W."*
2. The proposed conceptual design AKF 30304 dated 22 February 1937 is to be used as the basis for further development. The inside diameter of the turret ring is to be 1500 mm.
*3. The **7,5 cm Kw.K. L/24** (tank gun) from the **Pz.Kpfw.IV** was to be installed in the new turret. There were to be no changes allowed in the basic design of the gun with the exception of increasing the armor protection.*

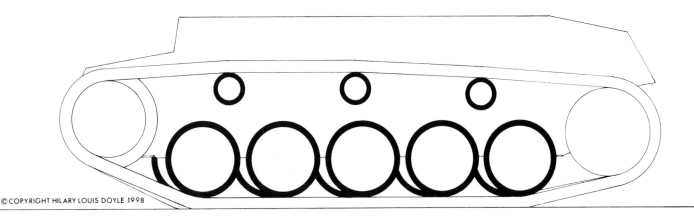

A sketch of the **D.W.2 Erprobungs-Fahrgestell** (experimental chassis) based on Dr. Aders' notes from February 1945 and one surviving photo.

4. All turret walls are to be 50 mm thick, the gun mantlet 20 mm, and the turret roof 15 mm.
5. A radio-receiver is to be mounted behind the gun in the turret.
*6. Due to the weight increase, the previous designs for the ball bearing turret race and gun elevating mechanisms are to be reviewed for acceptability. No plans are made for an electric drive for traversing the turret. Auxiliary traversing gear for the loader is to be included (standard in the **Pz.Kpfw.III**).*
7. The installation and design of all visors were discussed in Essen. Open bolt heads around the turret are to be avoided as much as possible.
8. It is intended to create an experimental turret directly from the plans, bypassing the intermediate stage of creating a wooden model.
9. Completion of the design and assembly is to be accelerated as much as the other development contracts allow.

On 24 June 1938, Wa Pruef 6/IId awarded Krupp contract Nr.106-8000/38 to complete a **D.W.-Versuchsturm** (trial turret) fabricated with cast iron. On 24 April 1939 Krupp was notified of the requested modifications to the **D.W.-Versuchsturm** in accordance with **Gruppenliste 021 Gr 15230**. Krupp had completed the **D.W.-Versuchsturm** by 26 May 1939 when they were ordered by Major Crohn, Wa Pruef 6/IId to ship the turret with its assembly stand to Grusonwerk, Magdeburg by 8 June 1939. It was planned to show a complete **Pz.Kpfw.IV** for comparison with other new developments, including the **DW-Versuchsturm**, to the head of the **Waffenamt** and other important persons.

1.2 VK 30.01

Following initial testing of the **D.W.** chassis, In 6 on 9 September 1938 authorized the **Heeres Waffenamt** to continue with the development of a Panzer in the 30 ton class. As discussed in a meeting with Krupp on 19 January 1939, Wa Pruef 6 stipulated that Panzers designed in the 30 metric ton weight class were to be developed only with the **7.5 cm Kw.K. L/24** and with the same crew space as in the **Pz.Kpfw.IV**. Heavier armament was not to be considered. Dr. Olbrich held the opinion that sufficient armor protection couldn't be achieved if larger caliber guns were mounted. With a 30 metric ton weight restriction, a Panzer could be designed with armor protection of 50 mm on both the front and sides. Armor plate 50 mm thick was the established standard for effective protection against uncapped armor-piercing shells fired by the German **3.7 cm Pak L/45** anti-tank gun.

Officially named the **Panzerkampfwagen VI (7.5 cm)** by 31 October 1940, it retained the design code name of **D.W.** along with the new code designation **VK 30.01**.

1.2.1 Henschel Fahrgestell

Henschel redesigned the **Fahrgestell**, now designated as the **VK 30.01**, incorporating lessons learned from the **D.W.1** and **D.W.2**. The new hull design, constructed as one piece, had 50 mm thick armor plates on the front, sides and rear, a 25 mm thick deck and 20 mm thick belly. The drive train consisted of a new six-cylinder **Maybach HL 116 Motor** delivering 300 metric HP at 3000 rpm, through a transmission onto the **Henschel L 320 C** steering gear and final drives. Maintaining the torsion bar suspension, the combat weight of 32 metric tons was distributed over seven sets of **geschachelte** (interleaved) 700 mm diameter roadwheels per side. The unlubricated 520 mm wide tracks had a single center guide, with the track pitch shortened to 160 mm.

Details of the hull armor were discussed in a meeting on 24 November 1939 between Baurat Rau of Wa Pruef 6 and Krupp representatives. Krupp had been awarded a contract to fabricate a target hull with superstructure based on the **VK 3001 alte Konstruktion** (old design) which was to be delivered to Kummersdorf by 23 April 1940 and subjected to armor penetration tests. In addition, Krupp had been awarded a contract to complete three **VK 30.01 neue Konstruktion** (new design) hulls with

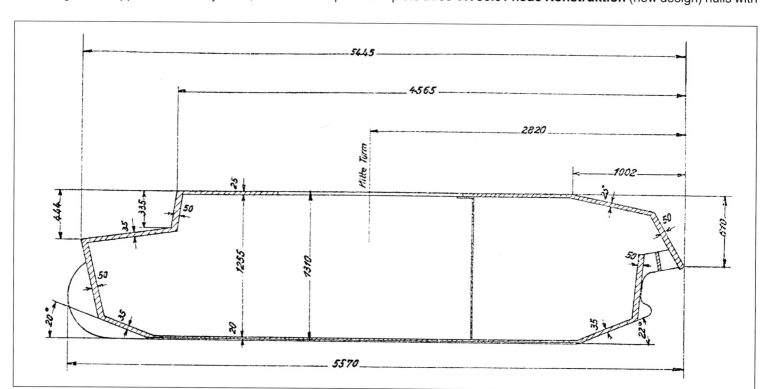

Dimensions of the hull with armor thicknesses from a Henschel drawing dated 17 January 1940

Chapter 1: Design and Development

superstructures in armor plate. One difference between the **neue Konstruktion** and the **alte Konstruktion** for the **VK 30.01** was that the vertical joint in the hull was dropped, with the new armor hull design being completed as one piece. Originally, **P 412** armor plate had been ordered for the target hull. The three new hulls were to be made from **PP 794** armor plate. Due to the urgency in the delivery schedule, Baurat Rau gave permission to create both the first **VK 30.01 neue Konstruktion** hull and the **VK 30.01 alte Konstruktion** target hull partially with **PP 794** and partially with **P 412** armor plates.

As shown in a drawing dated 17 January 1940, the target hull was redesigned to match the **VK 30.01 neue Konstruktion** with the following armor thicknesses and angles from vertical: 35 mm lower hull front at 70°, 50 mm hull front at 13°, 35 mm glacis at 32°, 50 mm driver's front plate at 9°, 15 mm thick deck at 90°, 25 mm rear deck at 79°, 50 mm upper hull rear at 30°, 50 mm hull rear at 5°, 35 mm lower hull rear 35 mm at 68°, and 20 mm belly at 90°.

This target hull wasn't completed by Krupp until after September 1940 for delivery to Kummersdorf to test its capability to provide complete protection against hits from 3.7 cm class weapons.

Details on the design of the **VK 30.01 Fahrgestell** were described by Dr. Aders in a report dated February 1945 as follows:

VK 30.01 of 30 metric tons with maximum speed of 35 km/hr
  Wanne (Hull) single piece. Crew entrance hatches forward on both the right and left side. Side extension for cooling air. Also experimented with cooling air intake by cutting rectangular slits across the rear deck over the engine compartment.
  Gleisketten (Tracks) unlubricated, 520 mm wide with one guide horn in the middle.
  Motor (Engine) six-cylinder Maybach HL 116 rated at 300 metric horsepower at 3000 rpm.
  Kuehlung (Engine Cooling) two radiators with four fans behind the engine belt driven with spring-tensioned rollers.
  Turm-Antrieb (Turret Drive) the same as in D.W.1 and D.W.2, with worm gears mounted on the main drive shaft. Insufficient lubrication led to unacceptable wear on the gears. Improved with a lubricating belt around the driving wormgear.
  Schaltgetriebe (Transmission) Maybach-Motorenwerk Variorex.
  Lenkgetriebe (Steering Gear) L 320 C Dreiradien (three stage) and hydraulic operation. Five clutches in oil bath.
  Seitliche Antriebswellen (Side Drive Shafts) the same as in D.W.1 and D.W.2, but both sides were the same length.
  Kettenantrieb (Final Drives) departure from the flanged shaft, turned to using the housing as the supporting body. Two planetary-gears in series with a gear reduction of 1 to 10.75.
  Bremsen (Brakes) Perrot interior-lined brakes largely adopted from the Z.W.38.
  Tragrollen (Return Rollers) three, the same as on D.W.1 and D.W.2, with rubber tires and ball-bearing mounting.
  Laufrad-Kurbeln (Roadwheel Support Arms) forged from a single piece because there wasn't room for the assembled style. Novotext bushings in the hull. The roadwheel arms were leading on the left side, and trailing on the right side.
  Stabfederung (Torsion Bars) simple torsion bars with adjustable heads (Porsche Patent!) for improved mounting and exact setting the support arms (Nonius-Effect). C value of 33 and 28 kg/mm.
  Laufraeder (Roadwheels) disc model with **Geschachtelte** (interleaved) alignment. Seven double-roadwheels per side.
  Stossdaempfer (Shock Absorbers) two forward and two rearward, directly mounted on the support arm shaft.
  Anschlaege fuer Laufrad-Kurbeln (Bump Stops for Roadwheel Support Arms) cast steel blocks with rubber cushions the same as on D.W.1 and D.W.2.
  Leitrad-Achsen (Idler Axle) adjustment inside with sheer pins and movement outward. Sealed for fording with curved covers made of armor steel.
  Leitraeder (Idler Wheels) steel casting with rubber tires.

Details of some changes introduced in the evolution of the **VK 30.01 Fahrgestell** design have survived in the following series of drawings:

Date	Drawing No.	Modification
12Oct39	021C39305 U5	**Laufrad** (roadwheel) 500 mm diameter
31Oct39	021D39305-9	**Stabfeder** (torsion bar)
29Dec39	VK 30.01	**Brennstoffbehaelter** (fuel tank) with 462 liter capacity
31Oct39	021B39311-4	**Zahnkranz** with 16-teeth spaced at 161 mm for 160 mm pitch track
5Jan40	021C39305 U5	700/98-550 **Laufrad Versteift** (reinforced roadwheel) 700 mm diameter
7Feb40	VK 30.01	**Brennstoffbehaelter** (fuel tank) with 408 liter capacity lost space due to relocation of torsion bars
29Feb40	021A39338	**Klappe** (hatch) for driver/radio operator
14Mar40	021B39329	**Hand u. Fusshebelwerk** (hand and foot operated controls)
21Jun40	VK 30.01	**Wanne** (armor hull) with 25 mm thick deck (all other thicknesses and angles remained the same as the drawing dated 17Jan40
18Jun40	021B39301 U1	**Decke, vordere vollst.** (complete superstructure) with oval hatch on side and 50 mm thick side armor
6Aug40	021B39311 U9	**Triebrad** (drive wheel) with 10 spokes and a 16-tooth sprocket ring with 940 mm outside diameter
11Oct40	021C39310 U1	**Stuetzrolle** (return roller) 340 mm diameter
27Nov40	021B39301-190	**Decke, vordere** (roof plate) with 1650 mm diameter hole cut out for the turret race
20Jan41	021D39324-40	**Doppel-Luefter** (double fans)
18Feb41	021C39325	**Auspuffanlage** (engine exhaust system)
25Feb41	021B39302 U1	**Getreibelagerung SSG 77**

Date	Part Number	Description
21Aug41	021B39320	(installation of **SSG 77** transmission) **Fusshebel f. 5.u.6. VK 30.01 der Nullserie** (foot pedals for the 5th and 6th 0-Serie VK 30.01 with **SRG 32 8 128** transmission)
26Aug41	021B39350	**Fusshebel f. 7.u.8. VK 30.01 der Nullserie** (foot pedals for the 7th and 8th 0-Serie VK 30.01) with **SMG 90** transmission)

Both the Maybach **HL 190** engine (375 metric horsepower) in 1935/38 and Maybach **HL 150** engine (400 metric horsepower) in 1938/40 had been proposed as alternative power plants to the Maybach **HL 116** engine. Plans were made to install and test several types of transmissions; including the 6-speed manual Zahnradfabrik **SSG 77** transmission, 10-speed semi-automatic Maybach SRG 32 8 128 transmission, and the 8-speed semi-automatic Maybach **SMG 90** transmission.

1.2.2 Krupp Turm

Krupp redesigned the turret complying with specifications from Wa Pruef 6 that with the exception of the armor thicknesses for the turret walls (50 mm for the **VK 30.01** and 80 mm for the **VK 65.01**), the turret for both the **VK 30.01** and the heavier **VK 65.01** were to be exactly the same. The design specifications called for a low profile turret with the gun centerline only 335 mm above the deck. The commander's hatch was to be surrounded by a cupola with a rotating ring mount housing seven periscopes and a **12-Uhr Zeiger** (azimuth indicator) ring. Additional vision devices in the turret consisted of three periscopes in the turret roof, two vision ports in the turret sides and the binocular **Turmzielfernrohr 9** (turret gun sight, model 9).

On 2 February 1939 the firm of Ernst Leitz GmbH, Wetzlar sent Krupp a drawing of a **T.Z.F.8** for the **D.W.** turret. Krupp reported to Wa Pruef 6/IId that the **T.Z.F.8** was too short for installation in the **D.W.** turret. On 21 February 1939, Krupp was informed that Wa Pruef 6 had arranged for delivery of a complete **T.Z.F.9**. With the exception of the reticle for the **7.5 cm Kw.K. L/24**, the **T.Z.F.9** sight was the same as the **T.Z.F.9b** installed in the **VK 45.01**.

On 28 May 1940, Wa Pruef 6/IId sent a letter requesting Krupp to design a machinegun mount firing to the rear: *For gun turrets it is required that a machinegun be fired to the rear independently of the machinegun mounted in the gun mantle. For the 7./B.W., VK 30.01, and VK 65.01 turrets, Wa Pruef 6 requests that Krupp look into a machinegun mount to be installed to fire to the rear with a similar design as previously planned for the 1.-3.Serie/Pz.Kpfw.IV. In regard to the thicker armor, attention should be paid to limiting the penetration to as small as possible with an armor hatch of sufficient mass to protect against hits.* On 19 June 1940, Krupp sent their proposal for a rearward firing machinegun port to Wa Pruef 6. The machinegun was fired through a small oval hole cut into the rear turret wall. A cylindrical armor guard bolted to the turret wall protected against penetrating hits. When not in use, an armor semi-disc was pivoted to seal the hole.

On 7 November 1940, Wa Pruef 6/IId requested that Krupp redesign the **D.W.** turret to install two **Turmbeobachtungsrohr T.B.R.1** (turret observation periscope) so that both the gunner and loader could see to the side.

As shown in drawing/part number list **021 Gr 39350** dated 11 September 1940, virtually all of the component parts had been designed specifically for the **VK 30.01 Turm (0-Serie)**:

021 St 39351 - **Beobachtungskuppel** (commander's cupola)
021 St 39352 - **12 Uhr Zeigerantrieb** (azimuth indicator drive)
021 St 39354 - **Sehschlitzplatte** (vision port)
021 St 39357 - **Lukendeckel, linker** (left turret roof hatch)
021 St 39358 - **Lukendeckel, rechter** (right turret roof hatch)
021 St 39360 - **Turmschwenkwerk** (turret traverse mechanism)
021 St 39362 - **Flussigkeitsgetriebe** (hydraulic drive)
021 St 39363 - **Hohenrichtmaschine** (elevation gear)
021 St 39367 - **Kommandantensitz** (commander's seat)
021 St 39368 - **Schuetzensitz** (gunner's seat)
021 St 39369 - **Ladesitz** (loader's seat)

The equipment list for the **VK 30.01** superstructure and turret dated October 1941 included:

1 **7.5 cm Kw.K.** (5 Gr 28)
2 **M.G.34**
29 **Patronengurtsack** (021 St 39150) (150 belted M.G.rounds)
1 **T.Z.F.9 bin** (027 Gr 185)
1 **K.F.F.2** (027 Gr 3539) (driver's twin periscopes)
1 **K.Z.F.2 in Kugelblende 50** (027 Gr 5075) (sight in ball mount)
1 **Schutzglas** 70x240x94 (021 St 9296) (driver's visor glass block)
2 **Schutzglas** 70x150x94 (021 St 9280) (vision port glass block)
7 **Winkelspeigel** (021 St 7621) (periscopes in commander's cupola)
1 **Aufbaulufter** (ventilation fan)
2 **Grundplatte EUa** (base mount for receiver set)
1 **Grundplatte U10a** (base mount for 10 watt sender set)
2 **Stabantenne (2 m)** (2 meter antenna, 1 spare)
1 **Nebelkerzenabwurfvorrichtung** (021 St 39160) (smoke candle rack)
1 **Panzerschutz f. N.K.A.V.** (021 St 39159) (armor guard on rear)
1 **M.P.** (machinepistol)
1 **Kurskreisel** (gyroscopic compass)
Behaelter (racks for 90-100 rounds of 7.5 cm ammunition)

A combat-ready **VK 30.01**, complete with turret, armament, and a crew of five, was estimated to weigh 32 metric tons.

1.2.3 VK 30.01 Production

On 24 November 1939, Krupp representatives met with Baurat Rau to discuss the production of three **Wannen VK 3001 neue Konstruktion** (hulls including superstructure) in armor plate but without the turrets. The **VK 30.01 neue Konstruktion** was different from the **alte Konstruktion** in that the vertical flange had been dropped and the new hull design was now completed as one piece. These three armor hulls were intended for **Fahrgestell** (chassis) to be used for driving trials. The first was to be delivered to Henschel in Kassel by 15 March, the second by 15 April, and the third by 15 May 1940. Krupp delivered all three armored hulls in 1940 for **Fahrgestell** assembly at Henschel. In February 1945, Dr.-Obering. Aders reported that the three **VK 30.01 Versuchs-Fahrgestelle** (trial chassis) had been completed and tested by Henschel.

Chapter 1: Design and Development

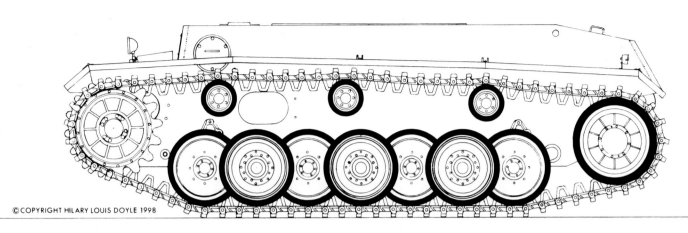

VK 30.01 Versuchs-Fahrgestell (trial chassis) **Fgst.Nr.V2** completed at the Henschel assembly plant by August 1941. A turret was not ordered for this chassis; it carried test weights instead.

On 23 July 1940, Wa Pruef 6/IId asked Krupp to deliver a complete **D.W.Turm** in armor, fully outfitted including hydraulic turret traverse. Contract 006-4489/40 was awarded to Krupp for this **D.W.Turm** on 14 October 1940 with the specification that it be completed in accordance with drawing 021 B 39350 out of **PP 793** armor plate (2.3% Cr, 0.3% Mo, 0% Ni, 0% W, & 0% Co). No record has been found stating that this single **VK 30.01 Versuchs-Turm** was completed.

On 29 January 1940, Wa Pruef 6 issued contract 106.3.3801/39 to Krupp, Essen to fabricate and deliver eight **0-Serie VK 30.01** armored hulls to Henschel from July to October 1941. Also on 29 January 1940, contract 106.3.5901/39 was issued by Wa Pruef 6 to Krupp-Grusonwerk, Magdeburg to assemble eight fully operational **0-Serie VK 30.01** turrets (the armored components for the turrets were to be supplied by Krupp, Essen under Wa Pruef 6 contract 106.3.5401/39 dated 15 February 1940. The operational turrets were to be delivered to Henschel in the period from October 1941 to January 1942 for mounting on their operational Fahrgestell in accordance with Henschel's contract 106.3.5902/39. During this same period, Henschel was awarded contract 106.3.5902/39 for superstructure assembly and contract 106.3.5001/39 for chassis assembly.

At a meeting on 18 September 1940 between Wa J Rue 6 and Krupp on steel works Nr.3 output: *In the discussion on assigning Apparatebau 3 a monthly production quota of 50 **B.W.** armor components, the question of delivery of VK 30.01 armor was introduced. Wa J Rue 6 stated that a contact for 40 **VK 30.01** armor components would be awarded to Krupp. These 40 armor components were to be delivered directly after the already contracted 8 for the **0-Serie** were delivered. Wa J Rue 6 couldn't give any further assurances that additional orders for **VK 30.01** armor components would be made, because the latest news was that the VK 30.01 was viewed as less urgent than other Panzer types currently being produced.*

On 17 April 1941, Herr Rau of Wa Pruef 6 requested that, if possible, the armor components for the 8 **VK 30.01** be delivered face-hardened. On 8 August 1941, Krupp sent the first **0-Serie VK 30.01** armored hull **Wanne Nr. 150411** to Henschel. The armored components for the first two turrets, **Turm Nr. 150411** and **150412**, were sent from Krupp, Essen to Krupp-Grusonwerk on 27 September 1941.

On 15 November 1941, Henschel, concerned about meeting the schedule for delivery of the first **0-Serie VK 30.01**, asked Krupp to deliver tools and equipment that were still needed to outfit the Panzer. This first **0-Serie VK 30.01**, with turret mounted, was located at the Henschel proving grounds in Sennelager and was to return to the factory toward the end of the month to complete outfitting. It was then to be delivered to the troops for testing the turret and combat equipment.

Krupp sent the last of the eight **0-Serie VK 30.01** armored hulls to Henschel on 30 November 1941 and the last armored turret body **Turm Nr. 150418** to Krupp-Grusonwerk on 21 January 1942. Maybach-Motorenwerk in Friedrichshafen produced three **HL 116** engines in 1940 and 11 in 1941. A further three **HL 116** engines were completed in 1942 and a final one in 1943.

On 30 January 1942, **Wa Pruef 6** reported that due to the substantial reduction in the workforce at Henschel, completion of the **0-Serie VK 30.01** by Henschel would provisionally be rescheduled. Of the eight **0-Serie VK 30.01**, only four were to be completed (under contract 106.3.5001/39 with two in March and two in April 1942), since they would be useful at a school for Panzer drivers. Completion of the other four **0-Serie VK 30.01** was to be delayed until further notice.

In a meeting on 25 September 1942, Oberstleutnant Kreckel, **Wa Pruef 6**, requested that Henschel complete four **Fahrgestelle VK 30.01** as **Schulfahrzeuge** (driver training vehicles) as quickly as possible. He hoped that these **Fahrgestelle** were close to completion, as Major Merker had furnished 15 men from his unit for this task.

At the end of 1942, Henschel reported that they had delivered a total of four **VK 30.01 Fahrgestelle** in 1942, two in March and two in October. Krupp-Grusonwerk reported that they had completed the assembly of four **VK 30.01** turrets in FY 1942 (1Oct41-30Sep42).

1.2.4 Attempts to Uparm the VK 30.01

On 7 October 1941 Krupp was asked by Wa Pruef 6 whether it was possible to replace the **7,5 cm Kw.K. L/24** with the **7,5 cm Kw.K. L/34,5** or another more effective gun. On 10 and 17 October 1941, Krupp replied that the **7,5 cm Kw.K. L/34,5** couldn't be mounted in the **VK 30.01** turret without extensive modifications. The **5 cm Kw.K. L/50** or **L/60** could be considered, because work on the design changes to the elevation mechanism and gun mantlet would be less. The tapered-bore **Waffe 0725** couldn't be installed.

In a further attempt to uparm the **VK 30.01**, on 3 December 1941 Wa Pruef 6 asked Krupp if the **7,5 cm Kw.K.44 L/43** de-

THIS PAGE AND OPPOSITE: One of the three **VK 30.01 Versuchsfahrgestelle** (trial chassis) which was used for trials with trench-digging equipment. (TTM)

Chapter 1: Design and Development

signed for the **Pz.Kpfw.IV** could be backfitted into the eight **VK 30.01** turrets, even if this resulted in greatly restricting the commander's actions. Krupp replied on 16 January 1942 that it was possible to mount the **7,5 cm Kw.K.44** but only with numerous design difficulties, including modifying the gun mount and designing a new elevation mechanism which would have required considerable time to complete. Wa Pruef 6 on 30 January 1942 decided to drop the rearmament project for the **VK 30.01** turrets.

1.2.5 VK 30.01 Turmstellung

The first indication that **VK 30.01** turrets were to be used for fixed emplacements was found in a notice dated 10 November 1942 on **Atlantikprogramm** drawings.

On 11 February 1944, Wa Pruef Fest reported on the status of Panzer turrets being used on stationary fronts. Six **Pz.Kpfw. VK 30.01** turrets had been acquired by Wa Pruef Fest from new production and were being modified for fixed emplacement by Krupp-Gruson, Magdeburg. Each turret was outfitted with a **7.5 cm Kw.K. L/24**, an **M.G.34**, a **T.Z.F.9** gunsight, and periscopes. One turret had already been installed in the West as a trial and additional turrets had been completed.

As reported by **Wa Pruef Fest** on 5 May 1944, six of the 0-Serie **VK 30.01** turrets with **7,5 cm Kw.K. L/24** had been released for emplacement in concrete stands as **Turmstellungen** (turret emplacements) along the **Atlantikwall**.

On 14 September 1944, In Fest reported that two **Pz.Kw.-Turm VK 30.01** left over from old issue for the **Atlantikwall** were available for installation as fixed defenses in the West. The two **Pz.Kw.-Turm VK 30.01** were issued to the **Hoeherer Pionier Kommandeur z.b.V.4** on 21 September 1944. **Heeresgruppe B** and **Hoeheres Kommando Saarpfalz** were requested to select the location and order their installation. The **Pz.Kw.-Turm VK 30.01** was to be installed in **Ringstand, Bauform 246**.

On 26 March 1945, all six **Pz.Kpfw. VK 30.01** turrets with **7,5 cm Kw.K. L/24** and an **M.G.34** were reported as having been installed in the **Atlantikwall** and **Westwall**.

1.3 VK 36.01

The history of the **VK 36.01** has always been shortchanged. In a summary status report dated 1 July 1942, the **Heeres Waffenamt** dismissed it as merely an interim step without bothering to mention the original project approval date or change in armament. Later, in February 1945, Dr. Aders of Henschel didn't even bother to describe the automotive evolution and characteristics of the **VK 36.01** separate from the **VK 45.01(H)** chassis to which it evolved.

Initially, in June 1940 the project was conceived as a Panzer in the 30 ton weight class mounting a turret with a 10.5 cm gun as its main armament. It was only following the decision to select a weapon with very high armor penetration capability on 26 May 1941, that the decision was made to drop the 10.5 cm gun turret and continue the **VK 36.01** project with a turret mounting the tapered-bore **Waffe 0725**.

Two innovations – cylindrical turret bodies and cast gun mantles – which would later influence another now famous turret design were initiated during the evolution of this **D.W. (10 cm)** turret.

The official name for this Panzer was revealed in an order from Wa Pruef 6/IIIf dated 21 October 1941 requesting that Henschel fill in the technical details in the acceptance specification to be used by **Heeres Waffenamt** inspectors examining completed **Pz.Kpfw.VI, Ausfuehrung B (VK 3601)**.

1.3.1 VK 36.01 (10.5 cm)

Already on 30 June 1939, Krupp had received a request from Wa Pruef 6 to design a turret for a new **Panzerkampfwagen** designated **A.W.**, the abbreviation for **Artilleriewagen**. The turret was to have 100 mm thick armor and mount a 10.5 cm L/20 to L/28 gun. Krupp completed the conceptual design by 20 October 1939. The 2.27 m wide turret mounting a 10.5 cm L/25 gun required a turret ring of 1.75 m diameter and weighed 8.4 metric tons. The entire **Panzerkampfwagen** was projected to weigh more than 80 metric tons.

On 5 July 1940, Dr. Olbrich of Wa Pruef 6 declared that the development of a Panzer with a 10 cm gun had gained in importance. Based on experience obtained during the recent campaign in the West, Panzers weighing over 30 tons would be of little value, since they were more or less restricted to major bridges. Considering this weight restriction, Krupp was to convert the **A.W.** turret design so that it could be mounted on a **D.W. Fahrgestell**, or convert the **D.W.** turret design to mount a 10.5 cm gun with the capabilities of the **l.F.H.18** (light field howitzer model 18). Krupp was given permission to halt further design work on the **A.W.** turret and was advised that it would probably be dropped.

Initially, the turret ring was to have a free inside diameter of 1.7 m. Frontal armor was specified to be 80 mm and side armor 50 mm. If weight restrictions allowed, a polygon-shaped turret was to be considered as well as a cylindrical design. In no case were periscopes to be mounted at roof level. Instead they must be installed in a special small protruding mount so that they wouldn't be threatened by every hit on the roof.

On 7 November 1940, Wa Pruef 6/IId requested that Krupp redesign the **D.W. (10 cm)** turret to install two **Turmbeobachtungsrohr T.B.R.1** (turret observation periscope) so that both the gunner and loader could see to the side.

In a meeting with on 21 November 1940, **Wa Pruef 6** asked Krupp to complete and deliver a **D.W.-Versuchsturm** (trial turret) in armor with two gun mantles by 1 July 1941. One gun mantlet was to be fabricated out of 80 mm rolled plate, and a second was to be made as an armor casting with a thickness providing equivalent protection as the 80 mm rolled plate. On 18 January 1941, Wa Pruef 6 awarded Krupp contract 006/4489/40 for a complete **D.W.Turm** with hydraulic traverse drive. Also, in January 1941, the previous contract 004/8027/40 awarded by Wa Pruef 4 on 16 September 1940 for the 10.5 cm L/28 gun to be mounted in the **A.W.Turm** was designated for use in the **D.W.Turm**. Krupp promised to deliver the gun in June 1941. On 19 March 1941, Wa Pruef 6/IId asked Krupp to send drawings of the **D.W. (VK 36.01) (Neukonstruktion)** (new design) to Henschel.

By 6 March 1941, Krupp had received notice that they were to deliver the armor hulls for four **VK 36.01 (Versuchsserie)**. They planned to deliver two armor hulls in January and two in February 1942. Contract SS-006-4086/40 for the assembly of four **D.W.Turm** in armor complete with armament was awarded by Wa Pruef 6 to Krupp on 5 May 1941.

Following the decision to mount a weapon with higher armor-penetration capability in late May, Krupp was notified on 11 June 1941 that the contract for one **D.W.Turm** with 10.5 cm L/28 armament was rescinded and the second contract for four turrets was to be converted and increased to six **D.W.Turm** with **Waffe 0725**.

1.3.2 Henschel Fahrgestell

In mid-1940, Wa Pruef 6 ordered Henschel to redesign the **D.W. Fahrgestell** to mount the new turret with a 10.5 cm gun. The weight of the resulting Panzer, designated **D.W. (VK 36.01)**, had now increased to 36 metric tons. Initially, the new hull design had 80 mm thick armor plates on the front, 50 mm on the sides and rear, and a 25 mm thick deck and belly. The drive train was upgraded to a new 12 cylinder **Maybach HL 174 Motor** delivering 450 metric HP at 3000 rpm, through an eight-speed **Maybach Olvar 40 12 16** transmission onto the **Henschel L 600 C** steering gear and final drives, designed to provide a maximum speed of 50 kilometers per hour. Retaining a torsion bar suspension, the weight of 36 metric tons was distributed over eight sets of **geschachtelte** (interleaved) 800 mm diameter roadwheels per side. The track width remained at 520 mm, but now had a pitch of 130 mm and two sets of guide horns.

Following the decision on 26 May 1941 to change the armament, Henschel proceeded to revise the design of the superstructure and internal layout of the **VK 36.01 Fahrgestell** to accommodate the new turret. With the ordered increase in the frontal armor thickness to 100 mm and side armor at 60 mm, the total weight of a complete combat-ready **VK 36.01** with a crew of five had increased to 40 metric tons.

A few details on changes introduced in the evolution of the **VK 36.01** design have survived in the following series of drawings:

Date	Drawing No.	Modification
18Mar41	021B4201-77	**Decke, vorn** (roof plate) 1830 mm wide with 1662 mm diameter hole cut out for the turret race
22Jul41	021B4201-101	**Decke, vorn** (roof plate) applicable starting with the second **VK 36.01**, 1830 mm wide and 25 mm thick with a 1766 mm diameter hole cut out for the turret race
14Jul41	021C4217	**Lenkgetriebe** (steering gear)
18Nov41	021C4228	**Lenkrad** (steering wheel)
17Nov42	021C4217	**Lenkgetriebe** (steering gear)
	021B4213	**Kettenglied** (track link) Kgs 63/520/130

At an engine speed of 3000 rpm the **VK 36.01** could achieve the following speeds in each gear with the **Olvar 40 12 16 Schaltgetriebe** (transmission invented by Maybach in 1940):

Gear	Ratio	Speed
Reverse	1:8.35	4.16 km/hr
1st	1:11	3.16 km/hr
2nd	1:7.38	4.72 km/hr
3rd	1:5.05	6.88 km/hr
4th	1:3.4	10.25 km/hr
5th	1:2.22	15.65 km/hr
6th	1:1.495	23.2 km/hr
7th	1:1.025	33.9 km/hr
8th	1.455:1	50.5 km/hr

1.3.3 VK 36.01 (Waffe 0725)

After reviewing the current status and plans for the development of Panzers and anti-tank weapons, Hitler made the following declarations on 26 May 1941:

- The development of both of the heavy Panzers from Dr. Porsche and Henschel is to be accelerated so that six of each will be available in the summer of 1942.
- The **7.5 cm Waffe 0725** (tapered bore gun) is acceptable for the Henschel design. However, this gun is to be produced in large numbers only if there is a satisfactory stockpile of tungsten. Approximately 1 kilogram of tungsten was needed for each high velocity, armor-piercing round fired by the **Waffe 0725**.

Chapter 1: Design and Development

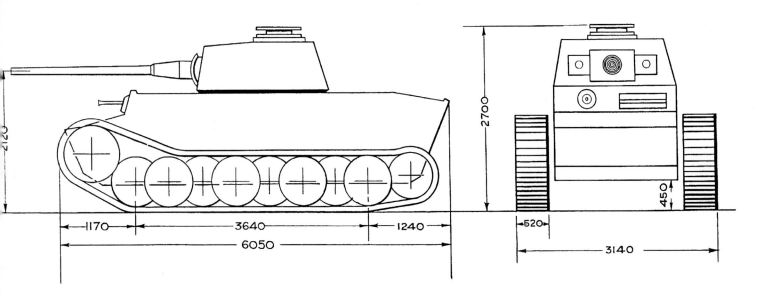

This sketch of a **VK 36.01** with a **7.5 cm Waffe 0725** mounted in the turret was traced from the original for a British Intelligence report. The original drawing probably dated from June/July 1941.

100 mm frontal armor thickness is considered necessary. 60 mm is sufficient for the sides of the Panzers to be completed by Porsche and Henschel in 1942.

In response to Hitler's decisions, on 28 May 1941 Wa Pruef 6 ordered Krupp to redesign the turret for the **VK 36.01** to mount the **Waffe 0725**. As shown in drawing Hm-C 455 dated 2 July 1941, the **Rohr 0725 fuer DW Turm** was 3.770 meters long (L/55.5 caliber length) and weighed 720 kilograms. This high-velocity gun was not fitted with a muzzle brake to reduce recoil.

The **7.5/5.0 cm Waffe 0725** was mounted coaxially with a **7.92 mm M.G.34** in an 80 mm thick curved gun mantlet. A rectangular armor vision port was installed in the 60 mm thick turret sides, and two machinegun ports were installed in the 60 mm thick turret rear. Hatches for both the gunner and the loader were cut into the turret roof. The commander's cupola, with seven periscopes in a rotating ring, was mounted in the left rear corner of the turret. Additional vision devices in the turret consisted of two periscopes in the turret roof for the loader and the binocular **Turmzielfernrohr 9b** for the gunner. With the exception of the reticle for the **Waffe 0725**, the **T.Z.F.9a** sight was the same as the **T.Z.F.9b** installed in the **VK 45.01**.

1.3.4 VK 36.01 Production

On 11 June 1941, Wa Pruef 6 informed Krupp that their contract for **D.W.** turrets would be revised to drop the **10.5 cm Kw.K. L/28** guns and the order increased to six turrets in armor plate. Henschel was awarded contracts to complete one **Versuchs-Fahrgestell** plus a **Versuchsserie** of six VK 36.01 Fahrgestell upon which six turrets from Krupp were to be mounted.

On 7 August 1941, Krupp's Abteilung AK sent out an internal memo advising all offices:

*The **VK 3601** is a forerunner of the **VK 4501(P)** and therefore is of very special priority. Since it belongs to the **Tiger-Programm** it must be given just as much priority as other projects under the program. The **VK 3601** is covered by the stamp: "**Fuehrerauftrag**" "**4501P**"(Tiger)*
Termine muessen unbedingt

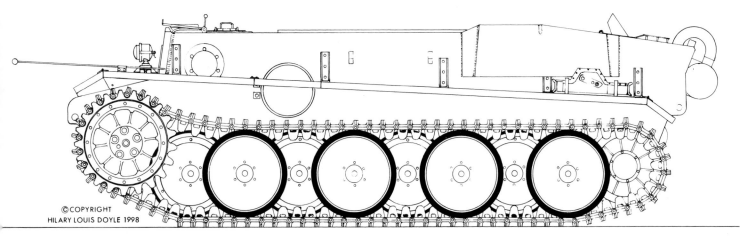

Henschel delivered the only **Versuchsserie VK 36.01 Fahrgestell** that was completed in March 1942. The turret designed and produced by Krupp was not mounted on the chassis.

eingehalten werden!
(Hitler's Order – All deadlines must be met unconditionally!)

It was established in July 1941 that an insufficient stockpile of tungsten existed to provide an adequate ammunition supply for a large number of **VK 36.01** with **Waffe 0725**. Therefore, in accordance with Hitler's directive the **VK 36.01** was not continued as a production series. Only the six **Versuchsserie VK 36.01** were to be completed.

A contract was awarded for the production of a single series of eight **7.5 cm Kw.K.42 (0725 in D.W.)** guns, to be delivered at the rate of two in November, three in December, and three in January 1942. On 5 December 1941, it was reported that due to start-up problems only one **7.5 cm Kw.K.42 (0725 in D.W.)** had been completed.

Wa Pruef 6 had altered contract SS-006-4086/40 for Krupp, Essen to manufacture and assemble six **VK 36.01** turrets with armament. On 20 January 1942, the production program laid out by Wa J Rue (WuG 6) called for the production of two **VK 36.01** turret armor shells per month starting in mid-February.

On 30 January 1942, Wa Pruef 6 reported that due to the substantial reduction in the workforce at Henschel, completion of the **VK 36.01** by Henschel would provisionally be rescheduled. Henschel was to complete two of the six **VK 36.01** (one under contract 006.6316/40 in March and one under contract 006.4080/40 in April 1942). Krupp had delivered four armored hulls and was ordered to deliver the final four hulls on an extended schedule. Krupp was to complete and deliver two turrets with **Waffe 0725**, but these were no longer urgently needed. On 10 February 1942, Krupp was ordered to complete assembly of the **VK 36.01** turrets but on a revised schedule that would not interfere with the **Tigerprogramm**. Dr. Aders of Henschel noted on 25 Februar 1942 that only two **Pz.Kpfw.VI Ausf.B (VK 36.01)** were to b assembled.

The status of **VK 36.01** turret assembly at Krupp was reporte on 17 March 1942 as:

a. Five **Kugellager** (ball-bearing turret races) are available in th shop. The sixth was rejected and a replacement will be deli ered.
b. Two **Geschuetze** (guns) are available in the shop. No reply t the question as to when further deliveries will occur.
c. We were informed by telephone that the gun sights won't b delivered because the turrets are to be used on chassis intende for driving trials.
d. According to the Waffenamt at a meeting on 5 February, th **M.G.** (machineguns) are to be installed at the ordnance depo Three machineguns needed for assembly will be made availabl
e. **Funk- und Bordsprechanlagen** (radio and intercom) hav yet to be provided by the Waffenamt.

Maybach-Motorenwerk reported that they had produced onl two **HL 174** engines in 1942. Krupp, Essen reported completio of eight **VK 36.01** armor hulls (one in FY41, seven in FY42) bu not a single turret in FY42 (1Oct41-30Sep42). Leitz didn't com plete or deliver any of the six **T.Z.F.9a** that were ordered in 1942

At the end of 1942, Henschel reported that they had deliv ered one **VK 36.01 Fahrgestell** in March 1942. This single V **36.01** chassis without turret was tested by Maybach i Friedrichshafen and took part in comparison trials at Berka i November 1942 (Refer to Appendix E). No evidence has bee found that Henschel completed additional chassis.

The only **VK 36.01** chassis completed by Henschel was sent down to Maybach in Friedrichshafen (Lake Constance in the background) for trials. (WJS)

Chapter 1: Design and Development

Driven by Reichsminister Albert Speer with Dr. Ferdinand Porsche on board, this **VK 36.01** chassis took part in the comparative trials with the Porsche **VK 45.01 (P)**, Henschel **VK 45.01 (H)**, M.A.N. **VK 30.01 (M)** and Daimler-Benz **VK 30.01 (D)** chassis in early November 1942. (WJS)

1.3.5 VK 35.02 (7.5 cm)

Consideration was also given to mounting a **7.5 cm Kw.K. (L/70)** gun (made famous in the Panther) in a turret mounted on the **VK 36.01** chassis. This was reported on 18 June 1942 as a **7.5 cm L/70** on the **VK 35.01** having the ability to penetrate 115 mm armor plate (set at 30 degrees at a range of 1000 meters with a 6.8 kg **Pzgr.39**). In a summary report on the status of projects under development dated 1 July 1942, the Heeres-Waffenamt reported that the **7.5 cm Kw.K.42 (L/70)** was being considered as the main armament in the **VK 45.01 (Henschel)**, **VK 30.02**, and a **VK 35.02**. A decision was made to stop pursuing the **7.5 cm Kw.K.42** option for the **VK 45.01H** on 14 July 1942, but no further mention in original records has been found on when the **VK 35.02** concept was dropped.

1.3.6 Schweres Abschleppfahrzeug (VKz 35.01)

On 20 June 1942, it was reported that expedient towing vehicles were to become available in November 1942 by utilizing five **Versuchs-Pz.Kpfw. VK 36.01** with **Seilwinde 22/40** (40 ton capacity winch).

In a summary report on the status of projects under development dated 1 July 1942, the **Heeres-Waffenamt** reported that Famo, Breslau had been ordered to design a **schweres Abschleppfahrzeug (VKz 35.01)** as a towing vehicle for heavy loads and recovery work. Maybach HL 210 engines, rated at 650 metric horsepower, were to propel the 35 ton fully tracked vehicle at speeds up to 35 kilometers per hour. Delivery of the four trial vehicles was estimated to occur in the spring of 1943.

On 25 September 1942, Oberstleutnant Kreckel of Wa Pruef 6 informed Henschel that in addition to the single **Versuchs-Fahrgestell** for Oberbaurat Kniepkamp, four **VK 36.01 Fahrgestell** were to be rapidly completed for towing Tigers. Known design work on the chassis still needed to be completed, because they were to be outfitted with a 40 ton winch from FAMO. A decision on the auxiliary drive from the transmission was especially needed. Furthermore, instead of the Maybach HL 174 engine as originally planned, the Maybach HL 210 (Tiger engine) was to be installed in these vehicles. According to Oberstleutnant Kreckel, Hitler had objected that there weren't any towing vehicles available for the Tiger. Therefore, completion of these four **VK 36.0 Fahrgestell** may not be neglected.

No **Heeres-Waffenamt**, Henschel, or FAMO records have been found proving that any of the **Schweres Abschlepp fahrzeug (VKz 35.01)** were completed and accepted for issue Nor have strength reports from any **Tiger-Abteilung** reveale that they had received any of these vehicles and used them a the front.

1.3.7 VK 36.01 Turmstellung

Eventually, the six **VK 36.01** turrets were ordered to be cor verted into **Turmstellungen** (turret emplacements) for installa tion in fixed defenses. The first indication that **VK 36.01** turret were to be used for fixed emplacements was found in a notic dated 10 November 1942 on **Atlantikprogramm** drawings.

Details on the conversion of **VK 36.01** turrets for Aktion Wa were reported by Krupp on 6 May 1943 as follows: *Rework th tapered-bore **Waffe 0725 (Rheinmetall)** for the **VK 36.01** to us ammunition produced for the **7.5 cm Pak 41 (Krupp)**. The cre in the turret should stand, therefore two seats can be delete The commander has a seat and new footrests are planned. H draulic traverse drive is dropped. The turret will only be traverse by hand. Foot pedals for firing the machinegun remain. The turr table supports are to be lengthened so that the gunsight is 145 mm above the platform. As a result of damage in March, part c the electrical equipment, periscopes, glass blocks, and other fi tings were destroyed. Lists of missing items are to be sent to th **Waffenamt** and reordered.*

On 11 February and 5 May 1944, Wa Pruef Fest reported o the status of Panzer turrets being used on stationary fronts. Si **Pz.Kpfw. VK 36.01** turrets had been acquired by Wa Pruef Fe: from new production and were being modified for fixed emplace ment by Krupp, Essen. Each turret was outfitted with a **7.5 cn Kw.K.** and an **M.G.34**, a **T.Z.F.9** gunsight and periscopes. Non of the turrets had been completed because of delays due to er emy bombing raids.

No **VK 36.01** turrets were reported as having been installe on any Front as of 25 March 1945. Photographs taken after th Allies occupied Essen show five of the six **VK 36.01** turrets still i various stages of completion at Krupp's steel works.

2
Panzerkampfwagen VI (Porsche)

In competition with Henschel, the firm of Dr.Ing.h.c.Porsche KG designed and developed a series of heavy and super-heavy Panzers from 1939 to 1944. Grave doubts were expressed by the Porsche organization as to the practicability of using a mechanical transmission for these heavy tanks. Therefore, Porsche used electric motors to drive his Panzers, similar to diesel-electric drives in rail locomotives, but with the power supplied by gasoline engines connected to electrical generators. Porsche designs were also unique in having torsion bar suspensions mounted externally to save space within the vehicle.

2.1 PORSCHE TYP 100

In late 1939, the firm of Porsche started their first attempt to design a Panzer, designated as **Typ 100**. While assisted by Wa Pruef 6, Porsche was given free rein to independently create their own design. Porsche concentrated on the design of the automotive components and let contracts to outside firms to produce other components, such as the hull armor by Krupp, air-cooled engines from Steyr, electrical components from Siemens. The contract for chassis assembly was awarded to Nibelungenwerk.

Porsche selected a gasoline-electric drive train for the **Typ 100**. Two air-cooled 10-cylinder engines were mounted beside each other in the rear of the chassis, each connected to an electric generator. The generated electricity was used to power two electrical motors mounted forward in the hull which in turn drove the front drive sprockets for the tracks. To save on internal space, the suspension consisted of three sets of paired roadwheels per side with each set sharing a longitudinally mounted torsion bar.

Having been awarded a contract to produce three armor hulls for the **VK 30.01 (P)**, Krupp contacted Porsche and offered to design a turret mounting an **8.8 cm Kw.K. L/56** gun in February 1941. In April 1941, Krupp was awarded contracts for the detailed design and production of six turrets with **8.8 cm Kw.K. L/56** guns for the **Typ 100**. These armor component contracts were superseded in July 1941 by contracts for **VK 45.01 (P)** armor.

A normal soft-steel hull for the **Typ 100** trial chassis was completed in July 1941 by another steel company in Austria, Eisenwerke Oberdonau. Two V-10 air-cooled **Typ 100** engines were also completed by Steyr in July 1941, tested by Porsche, and sent to Nibelungenwerk. All evidence points to only a single **Typ 100** chassis being completed and tested.

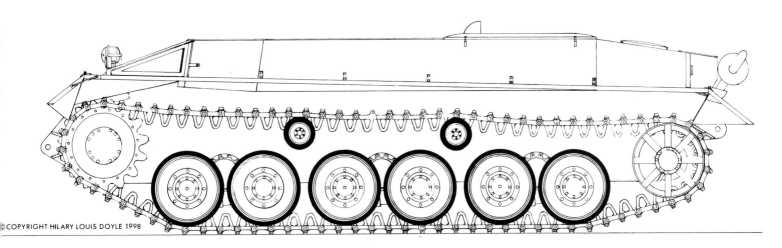

The Porsche **Typ 100** trial chassis completed at Nibelungenwerk using a soft steel hull produced at Eisenwerke Oberdonau.

THIS PAGE AND OPPOSITE: The only Porsche **Typ 100** trial chassis undergoing driving trials on a test track near Nibelungenwerk, where it was assembled. (WJS)

In a post-war report, Porsche stated that **Typ 100** trials had provided valuable information on electrical steering as well as the conditions for air-cooled engines in Panzers.

2.2 PORSCHE TYP 101

A meeting with Hitler on 26 May 1941 initiated the evolution of the Porsche **Typ 100** into the **Typ 101** design. It was decided to increase the frontal armor to 100 mm thick, retain the **8.8 cm Kw.K.** on the Porsche model, and pursue development of both Porsche and Henschel Panzers so that six of each would be available in the summer of 1942. The main difference between a Porsche **Typ 100** and **Typ 101** was the engine (10 liter **Typ 100** and 15 liter **Typ 101**), not the armament or weight classification.

The **Typ 101** project wasn't started until sometime in July 1941, as related by Dr. Porsche during postwar interrogation: *The decision was now taken to redesign the Leopard (**Typ 100**) as a larger tank to be called Tiger and the engine capacity was accordingly increased to 15 liters. The design was scarcely altered, all the main features of the 10-liter engine being retained. As a result of this the general arrangement drawings were in the hands of Simmering at the beginning of September 1941, in spite of the fact that the re-designing had only been started in July.*

Utilizing the **Typ 100** as its basis, Porsche did not redesign the **Typ 101** hull but merely changed details. When compared to a **Typ 100**, the **Typ 101** was upgraded with two air-cooled, V-10, Porsche **Typ 101** gasoline, 310 horsepower engines coupled to two Siemens 275 kilowatt generators. The drive train was moved to the rear of the hull and the motor compartment redesigned. Frontal armor was increased to 100 mm thick, side and rear armor to 80 mm thick. Steel-tired rubber-saving roadwheels replaced the rubber-tired roadwheels and the track return rollers were dropped. A 600 mm wide track was introduced to lower the ground pressure for cross-country travel. Additional data on specifications, dimensions, and capabilities are provided as Tables in Appendix A. Details on armor protection are included in Appendix D.

The turret with an **8.8 cm Kw.K. L/56** gun, originally designed by Krupp for the **Typ 100**, was mounted farther forward on the **Typ 101** chassis. Designed in a horseshoe shape (with the right side bending inward farther than the left), the opening between the 100 mm thick upper and lower turret front plates was protected by a cast gun mantle. Armor thicknesses were 80 mm for the turret walls and 25 mm for the turret roof. Details on armor protection are included in Appendix D. **Wa Pruef 6** assigned **Gruppen-Nummer 021 St 860** to the **VK 45.01 (P)** turret design. Drawing/part numbers for each component are listed in Appendix C.

Vision devices were installed to provide all-round observation by the crew in the turret. The gunner had a binocular **Turmzielfernrohr 9b** sighting telescope with 2.5x magnification and a vision block to his left. The loader had a vision block to the right front and a pistol port to the right rear. The commander had all-round vision blocks in the cupola and a pistol port to the left rear.

Access for the entire crew was provided by two hatches: one in the commander's cupola and a second in the turret roof directly over the loader's position. Secondary armament was provided by an **M.G.34** mounted coaxially to the right of the main gun.

In July 1941, Krupp was awarded contracts to provide 100 armor hulls for the **VK 45.01 (P)**. Krupp was also given the contract to assemble 100 turrets with armament in operational order. Both the armor hulls and operational turrets were to be delivered to Nibelungenwerk, which had been awarded the contract for chassis assembly.

ABOVE AND OPPOSITE TOP: Dr. Ferdinand Porsche on a visit to Nibelungenwerk in August 1942 where his **Typ 101** chassis were being assembled. This is one of the first eight **VK 45.01 (P)** which had the short-sided turrets designed and assembled by Krupp in Essen. (WJS)

©COPYRIGHT HILARY LOUIS

Chapter 2: Panzerkampfwagen VI (Porsche)

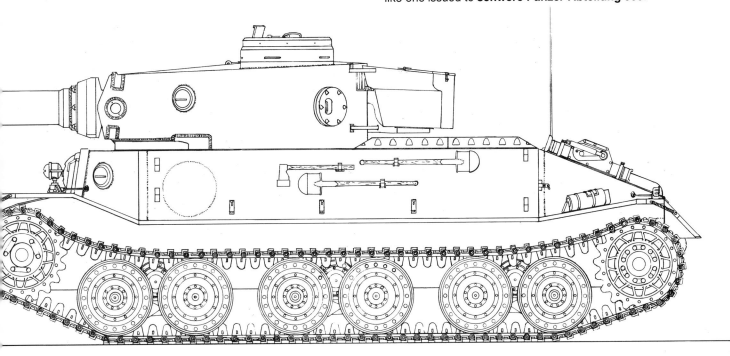

BELOW: A **Panzerkampfwagen VI P (Sd.Kfz.181)**, also known as the Porsche **Typ 101**, the **VK 45.01 (P)**, and the **Tiger (P)**, with tool stowage like one issued to **schwere Panzer-Abteilung 503**.

One of several **VK 45.01 (P)** with short sided turrets issued to **schwere Panzer-Abteilung 503** for troop testing and training. This **Tiger (P)** was purposefully driven through a marshy area to test its flotation. (WJS)

Initially, Nibelungenwerk was to have completed and delivered 10 production series **VK 45.01 (P)** in May 1942. Completion of one of these 10 was rushed to completion for display on Hitler's birthday in April 1942. Automotive problems disrupted production from the start. A single **Tiger P** was delivered for trials in June, none in July or August, one in September, and seven in October 1942. Total production achieved for the period from April to October 1942 amounted to only 10, compared to total production goals of 76 **Tiger(P)** for this same period.

The official end to **Panzerkampfwagen VI P** series production was decided at a conference with Hitler on 22 November 1942 when the proposal to produce 90 **Porsche Tiger I** with **8.8 L71** guns as **Sturmgeschuetz** was approved.

As with all production series German Panzers, modifications were frequently introduced during the production runs. In the case of the **Pz.Kpfw.VI P** these modifications were prompted mainly by a need to eliminate design faults so that a combat-serviceable Panzer could be fielded. The first eight turrets had lower sides and a flat roof with a raised center section to allow the gun to be depressed through a larger arc. The other 92 turrets for the **VK 45.01 (P1)** were exactly the same shape as mounted on the **VK 45.01 (H1)**; in fact 90 of the turrets originally ordered for the **VK 45.01 (P1)** were modified and mounted on the **VK 45.01 (H1)**. (Refer to Section 3.4.). A modified **Pz.Kpfw.IV** stowage bin was mounted on the turret rear and the **Pz.Kpfw.VI P** were outfitted with tools and equipment, starting in June 1942. A command version (**Befehlswagen**) of the **Tiger P** was also completed by Nibelungenwerk with the longer range radio sets and a **Rauchsignalkorb** (smoke signal basket) mounted in the dead space behind the commander's cupola. The **Nebelkerzenwurfgeraet** (smoke candle dischargers) still weren't available in early October, but the brackets were to be mounted and wiring run in the turret so that the **Wurfgeraet** could be mounted by the troops.

Out of the original order for 100 **VK 45.01 (P)**, 91 hulls were converted and completed as **Panzerjaeger "Tiger P"** with Fgst.Nr.150010 to 150100 (one by Alkett, 90 by Eisenwerke Oberdonau), three hulls were converted and completed as **Berge-Panzer VI** (recovery vehicles) and three **Fahrgestelle** were completed with **Ramm-Tiger** superstructures. As of 10 May 1943, four complete **Pz.Kpfw.VI P** with Turm Nr.150004, 150005, 150013, and 150014 were retained at Nibelungenwerk for further tests and trials. Subsequently, by August 1943, one of these was used for a **Rammtiger** chassis.

The only **Panzerkampfwagen VI P** known to have been used in combat was completed by Nibelungenwerk as a **Panzerbefehlswagen** with a new turret with higher roof. This **Befehls-Tiger** was taken to the Eastern Front with **schwere Heeres Panzer-Jaeger-Abteilung 653** in April 1944 and lost in July 1944.

2.3 Porsche Typ 102 Fahrgestell

As recorded in a postwar report: *When the Tiger tank project was started, grave doubts were expressed by the Porsche organization as to the practicability of using a mechanical transmission for so heavy a tank. Two alternative transmissions were therefore envisaged, one electric and the other hydraulic.* The Voith hydraulic transmission was coupled to the Porsche **Typ 101** engine. Each engine drove through its own hydraulic drive to a collector shaft and a combined forward/reverse and steering gearbox, which also accommodated an emergency low gear. As with the electric motors, the drive train in the **Typ 102** was designed to propel the **VK 45.01 (P)** at a maximum speed of 35 km/hr.

Original plans were to produce half of the 100 **VK 45.01 (P)** as **Typ 102** with hydraulic drives. But, due to delays in production the number was subsequently reduced to 10 and then canceled altogether. The actual reasons for the delay in delivering workable hydraulic drives on schedule was not revealed during postwar interrogations of Porsche or Voith. On 17 February 1943, Dr. Porsche reported that a **Tiger P1** with hydraulic drive was being completed at Nibelungenwerk. All evidence points to only a single **VK 45.01 (P) Typ 102** being completed and tested.

3
Panzerkampfwagen Tiger Ausf.E

The **Pz.Kpfw. Tiger Ausf.E** (also known as the **Tiger I**) was quickly designed, utilizing components that had been developed for and partially tested in previous heavy Panzers. Components for the chassis had been mainly invented for the 30 and 36 ton class of heavy Panzers in the **D.W.** series from Henschel & Sohn G.m.b.H., Kassel. The gun and turret for the **Tiger I** were designed by Fried.Krupp A.G., Essen for the competitor's 45 ton Panzer invented by Dr.ing.h.c.F. Porsche K.G., Stuttgart.

If any one of several key circumstances had been slightly altered, a completely different heavy Panzer would have been created instead of the now famous **Tiger I**. Among these key circumstances were:

- Problems with the automotive design for the Porsche-Tiger;
- Krupp's ability to maintain a monopoly on tank guns;
- An inadequate supply of tungsten for armor-piercing rounds;
- The report by Porsche that the higher performance **8.8 cm Flak** gun invented by **Rheinmetall** couldn't be mounted in the existing turret design; and
- The ease with which the previously designed **VK 36.01 Fahrgestell** could be modified to accommodate the larger turret already designed by Krupp for the **Porsche Typ 100**.

The following list of official designations is presented as an aid for keeping track of the names as they evolved during this design project:

VK 45.01 [28Jul41 - Henschel]

Pz.Kpfw.VI Ausf.H1 (VK 4501) [21Oct41 - Wa Pruef 6]

VK 4501 (H) [5Jan42 - Wa J Rue (WuG 6)]

Tiger H1 (VK 4501 - Aufbau fuer 8,8 cm Kw.K. Krupp-Turm) [Feb42 - Wa Pruef 6]

Pz.Kpfw.VI (VK 4501/H) Ausf.H1 (Tiger) [2Mar42 - Wa Pruef 6]

Pz.Kpfw."Tiger" (H) [20Jun42 - Wa J Rue (WuG 6)]

Pz.Kpfw.VI, VK 4501 (H), Tiger (H) Krupp-Turm mit 8.8 cm Kw.K. L/56 fuer Ausf.H1 [1Jul42 - Wa Pruef 6]

Panzerkampfwagen VI H (Sd.Kfz.182) [15Aug42 - KStN 1150d]

Tiger I [15Oct42 - Wa Pruef 6]

Pz.Kpfw.VI H Ausf.H1 (Tiger H1) [1Dec42 -]

Panzerkampfwagen VI H Ausf.H1 [Mar43 - D656/21+]
Corrected on cover to
Panzerkampfwagen Tiger Ausf.E

Pz.Kpfw.Tiger (8,8 cm L56) (Sd.Kfz.181) [5Mar43 - KStN 1176e]

Panzerkampfwagen Tiger Ausf.E (Sd.Kfz.181)
Panzerbefehlswagen Tiger Ausf.E [7Sep44 - D 656/22]

3.1 DEVELOPMENT

Development of the **VK 45.01 (H)** by Henschel was also initiated by decisions made during a meeting at the Berghof (Eagle's Nest) on 26 May 1941. After reviewing the current status and plans for the development of Panzers and anti-tank weapons, Hitler made the following declarations:

- The **7.5 cm Waffe 0725** (tapered bore gun) is acceptable for the Henschel design. However, this gun is to be produced in large numbers only if there is a satisfactory stockpile of tungsten.
- The feasibility of mounting the **8.8 cm Kw.K.** on the Henschel Panzer is to be investigated.
- 100 mm frontal armor thickness is considered necessary. 60 mm is sufficient for the sides of the Panzers to be completed by Porsche and Henschel in 1942.

In response to Hitler's decisions, on 28 May 1941 **Wa Pruef 6** awarded contracts for Henschel to redesign their chassis so

Chapter 3: Panzerkampfwagen Tiger Ausf.E

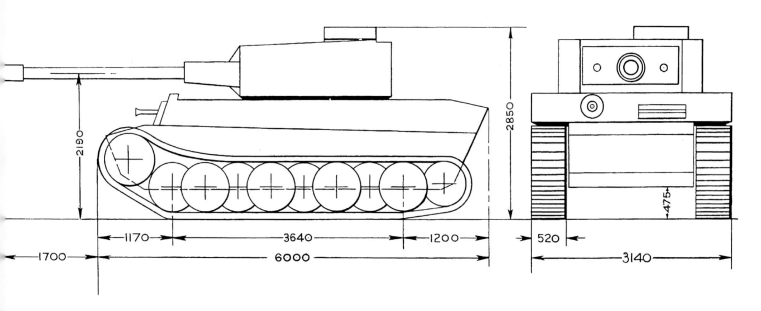

This sketch of a **VK 45.01 (H)** has several features that were changed during development, including a lower roof turret and an engine compartment similar to its predecessor the **VK 36.01**. Traced for a British Intelligence report, the original drawing probably dates from July 1941.

that it was suitable for mounting a turret with a **8.8 cm Kw.K.** and add armor protection for the tracks and drive sprockets.

As related by the head of Wa Pruef 6, Oberst Fichtner, on 27 September 1941: *Based on the further directive from Hitler in July 1941 that the tapered-bore guns should not be utilized, the turret developed by Krupp for Professor Porsche had to be taken over for the Henschel Panzer. Another solution was no longer possible because of the short time. This measure forced Henschel to modify their* **Fahrgestell** *so that it also became a 45 ton Panzer.*

The earliest drawing, found for the new chassis (HSK J2209 dated 28 July 1941), is entitled **Kuehlraum mit vergrosserten Kuehler** (cooling area with enlarged radiators). The superstructure sides had been extended out over the tracks to form a pannier in which to mount the larger radiators and cooling fans. These side panniers were not just adjacent to the engine compartment but extended forward to encompass the fighting compartment (later advantage was taken of this additional space for stowing 64 rounds of ammunition). Even at this early design stage, the angle iron supporting these large superstructure side panniers was to be welded to both the pannier and the hull (so that the superstructure could not be unbolted and removed from the hull, as had been the practice for the predecessor **Pz.Kpfw.I** through **Pz.Kpfw.IV** designs).

3.1.1 CHASSIS DEVELOPMENT

Dr.Ing. Aders, head of Henschel's design office, recalled in **Die Entstehung der Fahrzeuge Tiger E**, dated 6 February 1945, the first steps taken by Henschel in creating the **VK 45.01 (H)** chassis:

Important components (steering gear, final drives, suspension, idler, and drive sprocket wheels) could be borrowed from the **VK 36.01** *and used in mid-1941 as the basis for the design of the* **VK 45.01**.

Three weeks after the start of work on the design, the steel works could be informed of the program for the armor plates for the hull. After two months, the steel works received the fabrication drawings for the most important armor plates.

The design task was made especially difficult by special requirements for which solutions had to be found. Submerged fording in water up to 4.5 meters deep was required. An armor shield that could be raised and lowered was to protect the tracks against hits while driving on level ground.

After working through the conceptual design of the entire vehicle and determining the total weight and center of gravity, it was proven that the solid rubber tires for the roadwheels were insufficient for bearing the projected load of 58 tons. It was necessary to increase the number of rubber-tired roadwheels on each axle from two to three. Also, a design solution still had to be found for this.

New components that had to be created specifically for the **VK 45.01 (H)** were the:
- Engine cooling system located outside the engine compartment – two radiators with four fans protected by an impenetrable grating
- Watertight engine compartment deck
- Cooling system for the engine exhaust manifold that would also function to cool the transmission
- Turret drive taken off the main drive shaft
- Fuel system with four tanks, of which two had to be prepared for submersion under water 4.5 meters deep
- Air intake snorkel for submerged fording as a telescoping pipe
- Shock absorber arrangement and bump stops for the foremost and rearmost roadwheel arms
- Storage for 92 rounds of 8.8 cm ammunition (had to be designed twice, because the first design was created using unsuitable input)
- Holders for tools and equipment, inside and outside
- Installation of radio sets with antennae
- Hydraulically operated armor shield – raising and lowering device with controls and high pressure hydraulic pump system
- Bilge pump system for submerged fording

Components were developed by the following firms:

Engine	Maybach-Motorenbau
OG 40 12 16 transmission	Maybach-Motorenbau
Tracks	Ritscher-Moorburg
Brakes	Sudd. Argus-Werke

Turret and gun	Fried.Krupp
M.G. ball mount	Daimler-Benz A.G.
Driver's visor	Alkett-Berlin

Details on the design of the **VK 45.01 (H) Fahrgestell** were described by Dr.Ing. Aders in a report dated February 1945:

V.K 45.01 = Tiger I or Tiger E

Basic concepts established during a meeting at the end of June 1941 with Oberst Fichtner. Utilize components from the **VK 36.01** but with a new hull. Requirement for submerged fording capability was first raised.

Wanne (Hull): New with side panniers, because of the large turret ring diameter and also because side panniers were needed for the cooling system. Vertical hull sides, upper 80 mm thick, lower 60 mm. Connected to the pannier bottom plate with bolts and welding on angle iron made out of armor steel. Interlocking was rejected by Oberbaurat Rau. Reinforcement with a flat deck and a front plate angled at only 10° [sic]. In addition, the Alkett shutter visor designed for 80 mm wall thickness was adopted from the **VK 65.01** and **VK 36.01**.

An armor guard that could be raised and lowered was to protect the tracks against hits while driving on level ground. A high-pressure system consisting of hydraulic cylinders, geared supports, geared segments, levers, and shafts was necessary to raise and lower the armor guard. It was immediately rejected during the first demonstration for Hitler on 20 April 1942. Also, it wasn't sufficient to provide protection against shots fired from the side.

Gleisketten (Tracks): 130 mm pitch, 725 mm width. Originally different tracks for the right and left side. Later, only one model because of difficulty in supplying replacement parts. This caused a difference in resistance on both sides.

About 4 to 4.5 months after starting design work, it was learned that the total weight of the vehicle (55 instead of 40 to 45 metric tons) was too high for the load-bearing capability of the rubber tires. Another roadwheel was added on each axle and the track widened to 725 mm, instead of 520 mm as previously planned. Henschel proposed two tracks beside each other. **Wa Pruef 6** decided on a cross-country track and a loading track. The outer roadwheels were to be taken off for loading on rail cars.

Motor: HL 210 rated at 600-650 metric horsepower.

Kuehlung (Engine Cooling): For the first time (based on a proposal originating from M.A.N.) the radiators were located in side panniers beside the engine compartment. The radiators were flooded and the fans disconnected during submerged fording. Cooling air entered through the forward cast armor gratings, passed through the radiators, and was blown out by a total of four fans through cast armor S-shaped gratings on top at the rear. The fan housings mounted in pairs behind the radiators were curved to divert the air current. During high outside temperatures, the rear grating was to be raised and therefore mounted with hinges. It turned out that this wasn't necessary. The lifting device using hydraulics could be dropped.

Luefter-Anlage (Fan System): Two auxiliary drives were built into the rear of the Maybach HL 210 engine for the fan drives. Drive shafts with spur and bevel gears connected the engine to the fan drive mounted in the side bulkhead. Side shafts connected to a central drive in the fan housing, which drove two fans. Electromagnetic clutches were installed in order to disconnect the fans until the engine was warmed up, and also to avoid damage due to suddenly being over-revved.

Motorraum (Engine Compartment) was totally sealed for the first time. The only air circulation was from combustion air entering through an opening in the deck protected by armor **Hutze** (caps).

Because overheating from the glowing exhaust manifolds was foreseen, the exhaust manifolds were surrounded by ducts through which air flowed to cool them. The ducts were connected by flexible hoses to a box on the rear wall of the vehicle and sideward to the fan housings for the cooling system. Wa Pruef 6/III was of the opinion that the vacuum in the fan housing would be sufficient to move the cooling air. Henschel mounted a squirrel cage fan on front of the engine, which pulled cooling air through the enclosed transmission cowling and blew cooling air through the ducts surrounding the exhaust manifolds.

The **Auspuff-Anlage** (Exhaust System) consisted of two externally mounted vertical **Schalldaempfer** (noise mufflers), made out of sheet metal with internal dampers. The **Auspuff-Rohr** (exhaust pipes) penetrations through the vehicle tail plate were watertight. Large and strong cast armor guards protected the penetrations in the tail plate.

Tauch-Anlage (Diving System): Dives in water up to 4.5 meter deep were specified. The turret and hull roof had to be made watertight. All air intake openings and crew hatches, even the large cover plate over the engine compartment, were sealed with rubber gaskets. On the rear deck, a three-piece pipe was installed that could be put together and erected to create a three meter high snorkel as an air intake. Engine exhaust gases were blown out underwater; flooding of the exhaust mufflers was prevented by a hinged cap. Disconnecting the fans was achieved by the clutches, which also served as torque-limiting slip clutches.

Before diving, several butterfly valves in the ducts for the exhaust cooling air had to be closed or realigned to allow intake air to flow from the snorkel. A sump pump was set in operation to suck out any water that leaked in. Special heed was paid to airtight seals on the fire wall so that carbon monoxide gas couldn't endanger the crew (as learned from experience when diving with the Pz.Kpfw.III).

The **Turmantrieb** (Turret Traverse Drive), installed as a unit in the main drive shaft (just as in the **DW 1 and DW 2, SW (VK 65.01), VK 30.01**, and **VK 36.01**), had a ball clutch, a sump pump drive, and a high pressure hydraulic pump. All of these could be independently engaged and disconnected.

Schaltgetriebe (Transmission): Maybach-Motorenbau, Olvar 40 12 16 with hydraulic shifting after preselecting the gear. Maybach strived for extremely short gear-shifting times, independent of the driver's experience and training.

Lenkgetriebe L 600 C (Steering Gear) designed by Henschel. At first with three radii; then the smallest radius had to be dropped because of a weak point in the transmission. As in the **L 320 C** for the **VK 30.01**, shifting to the various radii was performed by liner clutches and hydraulic pressure. In case the hydraulic pressure failed, steering could still be done by using the driving brakes. The vehicle could be turned within its own length by disengaging the **L 600 C** steering gear.

Seitenwellen (Side Drive Shafts) the same as in **DW 1, DW 2, VK 30.01**: gear wheel-clutches with restricted movement.

Kettenantrieb (Final Drive) the same as **VK 30.01** with a gear reduction of 1 to 10.75.

Bremsen (Brakes): Disk brakes with a lining made out of synthetic rubber and steel shavings. Operated with a servo mechanism.

Chapter 3: Panzerkampfwagen Tiger Ausf.E

Gleisketten-Tragrollen (Track Return Rollers): Were totally dropped. Also, they were no longer possible because of the side panniers!

Laufrad-Kurbel (Roadwheel Arms): Made from one forged piece, adopted unchanged from the **VK 30.01**. Stress was very high (up to 50 kg/mm^2). Therefore, at first the arms attached to shock absorbers, then all of the roadwheel arms, were made of better quality steel.

Stabfedern (Torsion Bars): Complete cross cut, splined heads of various diameters for the purpose of setting the roadwheel arms at the same height. The front and rear torsion bars were somewhat stronger than the rest.

Laufraeder (Roadwheels): Disc wheels adopted from the **VK 36.01** with flat conical discs without waves, originally intended only for the first trials. Because they had significant experience with heavy **Zugmaschinen** design, Famo-Breslau was to develop new wheel designs. However, delays occurred when problems were revealed during cross-country trials with **Pz.Kpfw.** at Berka. Therefore, we had to settle on using the disc wheels because of the great rush to get the **VK 45.01** into mass production.

Bereifung (Tires): New type of hard rubber base furrows with wire inlays. Design still immature; very short life span of many tires causing the troops to constantly change the tires. This caused the **Schachtel-Laufwerk** (interleaved suspension) to be criticized.

Stossdaempfer (Shock Absorbers) adopted from the **VK 36.01**.

Anschlaege (Bump Stops) for roadwheel arms attached to shock absorbers, adopted from the **VK 36.01**.

Leitrad-Achsen (Idler Axles) the same as on the **VK 30.01**, but not designed to prevent breakage!

Leitraeder (Idler Wheel) copied from the design for the **VK 30.01**, but without rubber tires and with an armor steel hub.

3.1.2 KRUPP TURRET CONVERSION TO HYDRAULIC TRAVERSE

On 23 July 1941, Wa Pruef 6/IId awarded contract SS 006-6467/41 for Krupp to assemble three completely outfitted, operational **VK 45.01** turrets to be delivered to Henschel for mounting on **VK 45.01 (H) Fahrgestell** completed at Henschel under Wa Pruef 6/III contract SS 006-6307/41.

These turrets were exactly the same design which had been created by Krupp for the **VK 45.01 (P)**, with the exception that they were to be outfitted for hydraulic turret traverse drives on the Henschel chassis instead of electrical turret traverse drives on the Porsche chassis. Other minor changes made in the design for the **Krupp-Turm mit 8.8 cm Kw.K. L/56 fuer Ausf.H1** included the machine gun firing linkage, gun sight mount, azimuth indicator drive, equipment stowage, electrical layout, ventilation fan, and turret platform. Refer to Appendix C for a detailed list of drawing/part numbers that applied to each **VK 45.01** turret.

3.1.3 RHEINMETALL TURRET FOR 7.5 CM KW.K.42 L/70

An alternative turret was to be designed for the **VK 45.01**, as Oberst Fichtner, head of **Wa Pruef 6**, reported on 27 September 1942:

In a meeting on 25 July 1941 in Stuttgart-Zuffenhausen, I informed Prof.Dr. Porsche that I was not happy with the Krupp turret and strived for a better solution for the future that would be

The wooden model of the **Rheinmetall-Turm mit 7.5 cm Kw.K. L/70** (designated **Pz.Kpfw.VI H Ausf.H2**) with pistol ports on the turret sides, an emergency escape hatch on the right side, a communication port on the left side, and a machinegun ball mount on the rear. (APG)

*equally suitable for both the **Pz.Kpfw.Typ Porsche** and **Typ Henschel**. As already reported to Minister Dr. Todt, Wa Pruef 6 gave Rheinmetall a contract in mid-July 1941 to design a turret with a gun that can penetrate 140 mm thick armor at a range of 1000 meters without specifying that the caliber had to be 8.8 cm. The authority for this comes from Hitler's directive dated 26 May 1941, which states: "If the same penetration capability can be achieved by a smaller caliber than the 8.8 cm (i.e., 6 or 7.5), this should be given preference based on increased ammunition load and the lower turret weight. The chosen caliber must be suitable for engaging tanks, ground targets, and bunkers." Rheinmetall is attempting to achieve the penetration ability with a normal cylindrical gun tube based on the same principles as the **Pak 44**. It would have been wrong to pass up this Rheinmetall project because in meeting all the necessary requirements (rate of fire, ammunition load, balanced turret weight, observation conditions), it foreseeably leads to a more advantageous turret than the current Krupp turret.*

*The first **7.5 cm Kw.K. Versuchsrohr** (trial gun tube) of L/60 caliber length, designed and test fired by Rheinmetall-Borsig, just* met the specified penetration ability of 100 mm thick armor plate at 30° at 1400 meters range. Therefore to ensure that the penetration specification was met, a final gun tube of L/70 caliber length was chosen. By 11 February 1942, Rheinmetall had designed the **VK 45.01 (Rh)** turret with a **7.5 cm Kw.K.42** to be mounted on the **VK 45.01 (H)** chassis. The **VK 45.01 (H)** with **Rheinmetall-Turm mit 7.5 cm Kw.K. L/70** had been officially designated as the **Pz.Kpfw.VI H Ausf.H2** by **Wa Pruef 6** by 1 July 1942.

On 1 July 1942, **Wa J Rue (WuG 6)** revealed long range plans under **Hitler Panzerprogramm II** to produce only the first 100 production series **VK 45.01 (H)** with the **8.8 cm Kw.K. L/56**. Then starting with the 101st **VK 45.01 (H)** in February 1943 production was to be shifted to the Rheinmetall turret with the **7.5 cm Kw.K. L/70**.

The subject of Tiger armament was discussed at a **Panzerkommission** meeting on 14 July 1942: *Recently the ability to penetrate 100 mm of armor was also achieved with the **8.8 cm Kw.K. L/56**, therefore conversion to the **7.5 cm Kw.K. L/70** is no longer necessary. Conversion to the **8.8 cm Kw.K. L/71** should occur at the end of this year.* This decision resulted in the entire production run of **VK 45.01 (H)** being outfitted with turrets mounting the **8.8 cm Kw.K. L/56**.

There wasn't any change to the **8.8 cm Kw.K. L/56** gun itself, just to the design of the armor-piercing shells. Greater armor penetration was achieved by decreasing the size of the explosive filler cavity inside the shell which also slightly increased its weight to 10.2 kilograms. (Refer to Chapter 7 for more details on armor penetration capability.)

3.2 DESCRIPTION

3.2.1 FAHRGESTELL

The following detailed description and illustrations of the chassis designed by Henschel were extracted from the **Vorlaeufige kurze Beschreibung des Panzerkampfwagen VI H Ausfuehrung H1** dated 1942 and **D656/21+ Panzerkampfwagen VI H Ausfuehrung H1 Firmen-Geraetbeschreibung und Bedienungsanweisung zum Fahrgestell** dated March 1943.

The **Fahrgestell** (chassis) consists of the following components:
Panzerwanne (Armor Hull)
Maybach HL 210 P45 (Engine and Accessories)
Turmantrieb (Turret Drive)
Maybach OG 40 12 16 Schaltgetriebe (Transmission)
L 600 Zweiradien-Lenkgetriebe (Steering Gear)
Bremsen (Brakes)
Seitenvorgelege (Final Drives)
Schachtellaufwerk (Interleaved Suspension)
Gleisketten (Tracks)
Kraftstoffbehaelter (Fuel Tanks)
Kuehler- und Luefteranlage (Radiators and Cooling System)
Beluftung (Ventilation)
Lenzanlage (Bilge Pump)
Elektrische Anlage (Electrical System)
Selbsttaetige Feuerloescheinrichtung (Automatic Fire Extinguisher System)
Munitionslagerung (Ammunition Stowage)
Zubehoer (Tools and Equipment)

3.2.1.1 Panzerwanne (Armor Hull)

The 100 mm thick driver's front plate is angled at 9 degrees from vertical, 100 mm front nose plate at 25 degrees, 60 mm glacis plates at 80 degrees, 80 mm superstructure side plates 0 degrees, 25 mm pannier floor horizontal, 60 mm hull side plate at 0 degrees from vertical, 80 mm tail plate at 9 degrees, 25 mm deck plates at 90 degrees horizontal, and 25 mm belly plate horizontal. Refer to Appendix D for further details on armor specifications.

The armor hull is divided into the fighting compartment, the closed engine compartment, and two side panniers open at the top. The turret drive, transmission, steering gear, brakes, place for the driver on the left and radio operator on the right, and the entire ammunition stowage for the main gun and machinegun are located in the fighting compartment. The engine with inertia starter, engine tool box, mechanical fuel pump, the two lower fuel tanks, rear shock absorbers, and fan drive are located in the engine compartment. An upper fuel tank, radiator, and fan are located in each of the side panniers to the right and left of the engine compartment.

A large round opening in the middle of the armor deck over the fighting compartment is left free for the turret. The traversable turret with main gun and machinegun sits in this hole, mounted on a ball-bearing race. The hydraulic drive for traversing the turret is mounted on the turret platform which hangs below the turret down in the hull.

Two crew hatches with closable lids are located to the right and left in front of the turret on the armor deck. A driver's visor and a machinegun ball mount are fitted into the driver's front plate.

Openings in the bottom of the hull include:

a drain cock in the front right by the radio operator's seat, a drain cock at the back to the left of the drive shaft, a cover plate toward the front left under the steering gear, a cover plate toward the front center under the transmission, a drain cock on the right side in the engine compartment, and two cover plates under the engine.

Towing shackles, mounted in holes cut into hull side extensions at the front and rear, are for connecting tow cables in either horizontal or vertical axis or for connecting rigid towing bars. In addition, a tow coupling is welded to the hull rear for towing **Kraftstoffanhaenger** (fuel trailer).

3.2.1.1.1 Lukendeckel (Driver's and Radio Operator's Hatches)

Two round hatches are located above the driver and radio operator in the roof of the hull. The hatch lids are hinged on a connecting band so that they can be swung outward. Inside each hatch lid are three closure bolts that are operated by a lever pivoting from the middle of the hatch lid. In addition, the hatch lid can be closed watertight with three tensioning screws.

Two spiral springs attached to the connecting band balance the weight of the hatch lid so that very little effort is needed to open or close the lids. When the hatch lid is opened all the way, a spring-tensioned hasp engages a notch in the connecting band to prevent the lid from unintentionally slamming shut.

A periscope is mounted in each of the hatch lids to widen the driver's and radio operator's field of view. An armor guard over the periscopes protects them against attack by strafing aircraft.

3.2.1.1.2 Fahrersehklappe (Driver's Visor)

The driver's visor mounted in the driver's front plate has an external double sliding armor shutter that can be opened and closed with a handwheel. When the visor is open, the driver looks through a 70 mm x 240 mm x 94 mm laminated glass block. When the visor is closed, the driver uses the **K.F.F.2** twin periscopes whose heads are aligned with twin holes cut into the driver's front plate. The **K.F.F.2** has a field of view of 65 degrees and a magnification of 1x.

3.2.1.1.3 Kugelblende 100 (Machinegun Ball Mount)

The machinegun ball mount is inset into the driver's front plate in front of the radio operator. An external armor guard and a fitting ring cover the joint between the ball mount and the driver's front plate. The **M.G.34 mit Panzermantel** (machinegun model 34 with armor barrel sleeve) and the **Kugelzielfernrohr 2** gun sight are mounted on the rear of the ball. The ball mount can traverse 15 degrees to the right and 15 degrees to the left of center and can be elevated through an arc of -10 to +20 degrees. The **K.Z.F.2** has a field of view of 18 degrees and a magnification of 1.8x.

3.2.1.2 Maybach HL 210 P45 (Engine and Accessories)

The Maybach 21 liter, V-12 gasoline **HL 210 P45** engine delivers 650 metric horsepower at 3000 rpm. The engine is equipped with three Mahle vortex air filters, four Solex twin-jet carburetors, two Schnapper magnetos, four mechanical fuel pumps, an electrical fuel pump, an electrical starter, an inertia starter, a 1000/12 generator, a water-cooled oil cooler, and an automatic fire extinguisher system. A governor, controlling the butterfly valve between the fuel intake and carburetor independent of the gas pedal linkage, prevents engine overspeed.

3.2.1.3 Turmantrieb (Turret Drive)

The torque is transmitted from the engine to the transmission by the main drive shaft, split in the middle by an intermediate bearing for the turret drive. The gears of the turret drive are operated by an auxiliary drive shaft from the transmission. The connection is made through a conical coupling which is operated from the fighting compartment. This installation permits the turret to be traversed as needed by hydraulic power. The auxiliary drive shaft also powers the bilge pump.

3.2.1.4 Maybach OG 40 12 16 Schaltgetriebe (Transmission)

The Maybach **Olvar 40 12 16** transmission with built-in clutch operates semi-automatically. The direction lever is manually operated and has three positions: forward, neutral, and reverse. The eight forward and four reverse speeds are selected by a small lever mounted near the steering wheel. Depressing the selector lever and the clutch pedal activates the gear change, which is actually made by hydraulically operated cylinders built into the transmission.

Gear ratios and the maximum vehicle speed obtainable in each gear at an engine speed of 3000 rpm were:

Gear	Ratio	Speed
Reverse	1:8.35	3.7 km/hr
1st	1:11	2.8
2nd	1:7.38	4.2
3rd	1:5.05	6.1
4th	1:3.4	9.1
5th	1:2.22	13.9
6th	1:1.495	21.3
7th	1:1.025	30.0
8th	1.455:1	45.0

3.2.1.5 L 600 Zweiradien-Lenkgetriebe (Steering Gear)

The steering mechanism consists of a bevel drive and two symmetrically centered planetary gears which transmit the power from the transmission to the final drive shafts. The radius of the turning circle is dependent on the transmission gear and steering clutches engaged:

	Steering Clutch	
Gear	Smaller	Larger
Reverse	4.7 m	14.3 m
1st	3.57	10.85
2nd	5.45	16.5
3rd	7.75	23.5
4th	11.5	35
5th	17.7	53.7
6th	26.2	79.7
7th	38.3	116
8th	57	173

If the main clutch in the transmission is disengaged when the vehicle is stationary, the vehicle turns on the spot as one track pulls forward and the other track pulls rearward (turning speed can be increased by revving up the engine).

3.2.1.6 Bremsen (Brakes)

The brakes are mounted on the inside of the Panzer on the left and right sides of the final drive shaft. The brakes can be applied to both tracks at the same time by using a foot pedal or individually by using one of the two hand-operated steering levers. Thus, the Panzer can be steered by selectively applying the right or left brake by using the respective hand lever. A parking brake is provided by a locking mechanism on a hand lever.

3.2.1.7 Seitenvorgelege (Final Drives)

Power is transmitted from the steering mechanism to both final drives, one on each side. Each final drive consists of shafts, one spur gear countershaft, one planetary gear, and one driving wheel with two sprockets for driving the tracks.

3.2.1.8 Schachtellaufwerk (Interleaved Suspension)

Because of its weight, the suspension for this Panzer had to be designed as a **Schachtellaufwerk** with interleaved road wheels. There are eight torsion arms on each side, leading on the left side, and trailing on the right side. Each of the torsion arms is attached to a torsion bar mounted transversely across the bottom of the hull. Three roadwheels are supported by each torsion arm. The outer roadwheel is connected by a bolted flange so that it can be removed to reduce the overall vehicle width for loading on a railcar. In traveling over obstacles, the up-and-down motion of the torsion arm is dampened by the torsion bar. Single-acting, hydraulic shock absorbers are attached internally to the two front and two rear torsion arms. The movement of these four torsion arms is limited by rubber bump stops.

3.2.1.9 Gleisketten (Tracks)

Because of the Panzer's weight, two types of tracks – **Marschkette** and **Verladekette** (operational and transport tracks) – were needed to achieve the lowest possible ground pressure. Ground pressure of 1.11 kg/cm^2 is achieved with the 725 mm wide **Marschkette**, consisting of 96 unlubricated track links per side. The outer roadwheels are removed and the **Marschkette** is replaced by the **Verladekette** before loading the Panzer on rail cars. The ground pressure with the narrower **Verladekette** (520 mm wide) increases to 1.545 kg/mm^2.

3.2.1.10 Kraftstoffbehaelter (Fuel Tanks)

A total of 348 liters of fuel is carried in the two upper fuel tanks (mounted in the side compartments above the engine) and 186 liters in the two lower fuel tanks (mounted to the left and right of the engine). An 8 mm thick armor plate protects the upper fuel tanks from any shell fragments or small caliber rounds that may penetrate the armor grill bolted to the hull above the fuel tanks. Also, a wire mesh screen in a frame is bolted to the outside of the armor grill to prevent explosive charges from being thrown onto the fuel tanks through the slits in the armor grill.

3.2.1.11 Kuehler- und Luefteranlage (Radiators and Cooling System)

The armor grill also serves as the air intake for cooling air flowing through the radiators, which are located directly behind the fuel tanks. One radiator and two fans are located on each side.

The fan drive housing is connected to the engine compartment side wall and sealed watertight with a rubber ring. The fan area is covered with a heavy armor grill through which the radiator cooling air is exhausted. A wire mesh screen in a frame is bolted to the outside of the armor grill as protection against hand grenades.

During submerged fording the side compartment, in which the upper fuel tanks, radiators and fans are located, is sealed off from the engine compartment and flooded. The water drains out through holes in the hull rear behind the engine exhaust mufflers.

3.2.1.12 Belueftung (Ventilation)

A fan is built into the firewall between the fighting and engine compartments for the purpose of cooling the transmission. Air, drawn out of the shroud surrounding the transmission, is pulled through a duct running along the hull floor and blown into the jackets surrounding the engine exhaust headers. From there, the warm air is directed through ducts penetrating the side compartment walls into the fan housing and exhausted.

For submerged fording, the engine compartment air intake vent is sealed and a snorkel is erected on the rear deck through which fresh air is drawn into the engine and fighting compartments.

3.2.1.13 Lenzanlage (Bilge Pump)

The bilge pump system is used to pump out the water that seeps into the Panzer when traveling submerged. The centrifugal suction pump (rated at 250 liters per minute) is connected to a pipe which discharges the bilge water outside over the fuel tank on the right side. The end of the discharge pipe is covered with a hinged flap to prevent water from flowing backward down the pipe into the Panzer when it is submerged.

The bilge pump is operated by using the double linkage control, which first starts the turret drive and then engages the bilge pump through a dog clutch. Attached to the discharge pipe is a funnel with a valve through which water and antifreeze solution can be poured into the pump so that it is ready for action even in freezing weather.

3.2.1.14 Elektrische Anlage (Electrical System)

The electrical installation consists of the lighting system and the ignition system. A twin polar main battery switch built into the firewall divides the entire system from the source of current – two 12 volt, 150 amp-hour storage batteries. As controlled by a regulator, the batteries are charged by a generator when the Maybach HL 120 P45 engine is running.

The lighting system consists of two instrument panel lights, one radio operator light, two dismountable headlights, one waterproof tail light, and the turret lighting.

The radios and intercom are connected to the 12 volt battery in such a way that the main power switch cuts out the positive line. Four suppressors, one for the fuel pump and three for the regulator, are installed to eliminate radio interference; otherwise the wires are double insulated.

3.2.1.15 Selbsttaetige Feuerloescheinrichtung (Automatic Fire Extinguisher System)

The fire extinguisher system, installed in the engine compartment, is directed against the carburetors and fuel pumps. The extinguishing system is fully automatic, using CB as the extinguishing agent. When the thermostats by the spray heads sense a temperature exceeding 120°C, a discharge valve is opened for 7 seconds. If the fire is not overcome in 7 seconds, the next discharge is started immediately. The extinguisher flask holds 3 liters, which is sufficient for five discharges.

3.2.1.16 Munitionslagerung (Ammunition Stowage)

Ammunition bins are fitted with metal lids which are to be kept closed. Sixty-four 8.8 cm rounds are stowed in 16-round bins in the panniers alongside the fighting compartment. Six rounds are stowed as a reserve in a bin beside the driver. These 70 positions are long enough to stow either armor-piercing or high-explosive shells.

Sixteen armor-piercing shells are stowed in ammunition bins along the hull sides below the panniers. A reserve of six armor-piercing shells are stowed in a bin below the turret platform. These six shells are accessible through a hatch in the turntable floor.

3.2.1.17 Halterungen mit Zubehor aussen am Pz.Kpfw. (External Stowage)

Chapter 3: Panzerkampfwagen Tiger Ausf.E

Tools and stowage locations changed continuously during the Tiger I production run (Refer to Section 3.4.1.4). The following items were extracted from an **Ausruestungsliste fuer Pz.Kpfw.VI H, Ausfuehrung H1**, effective February 1942:

- 1 tool box for track tools
- 1 stowage bin (on turret rear) for 10 spare track links and pins and a 1200 mm x 1800 mm tarp
- 1 15-ton steel jack with lifting housing and folding crank
- 1 centering guide for inertia starter
- 1 hand crank shaft for inertia starter
- 1 14 mm diameter steel cable, 15 meters long
- 1 2-liter Tetra fire extinguisher
- 1 axe
- 1 sledge hammer, 6 kg
- 1 wire cutter
- 1 wooden block for the 15-ton jack
- 2 32 mm diameter steel cable, 8.2 meters long
- 6 gun cleaning rods (five 980 mm, one 918 mm long)
- 1 shovel
- 1 spade
- 1 crowbar, 1800 mm long
- 1 plate to cover air intake slit on the engine hatch
- 4 tow shackles

GERMANY'S TIGER TANKS - D.W. to TIGER I

Panzerkampfwagen VI H Ausf.H1 – Fgst.Nr.V1 completed in April 1942 – **Vorpanzer** with extended glacis side plates – horn on superstructure roof – no periscopes on driver's or radio operator's hatches – no air intake in engine hatch – antenna base on hull rear – cupola hatch lid opens to lay flat

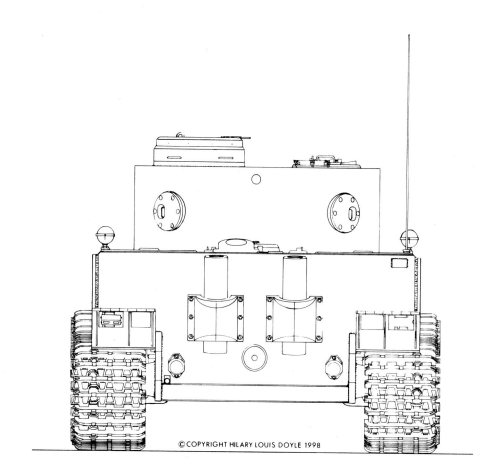

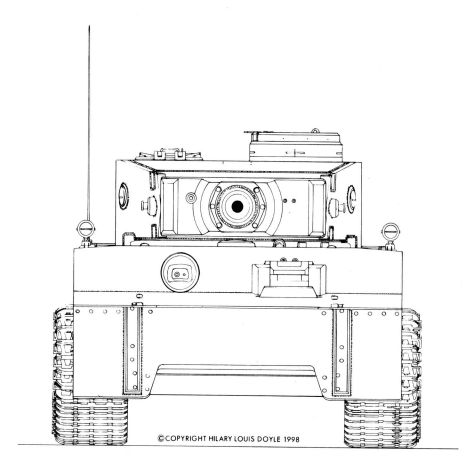

Chapter 3: Panzerkampfwagen Tiger Ausf.E

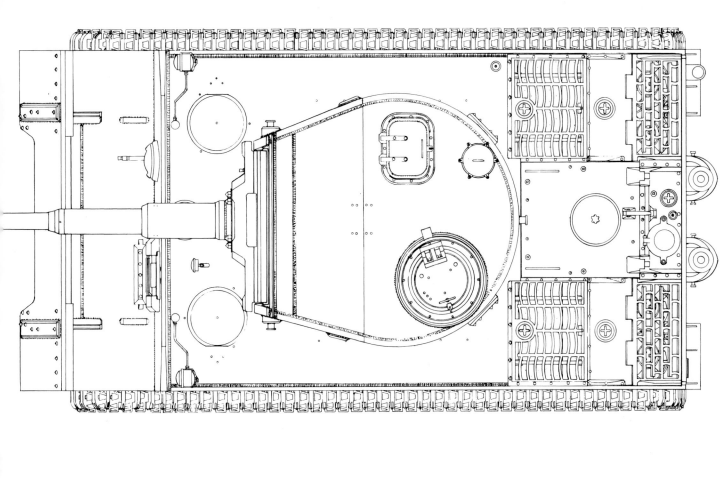

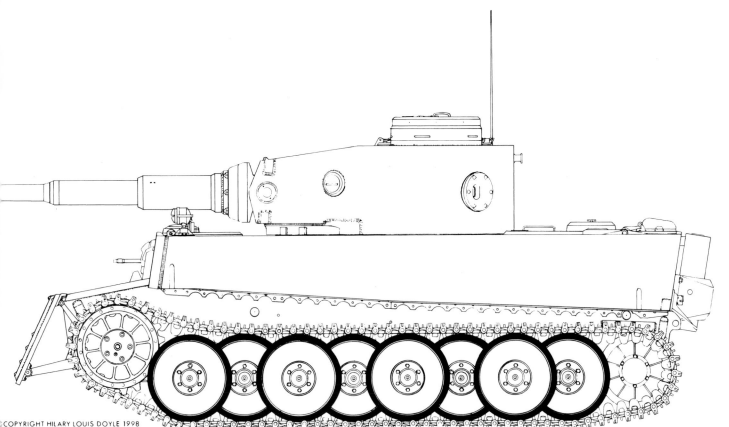

39

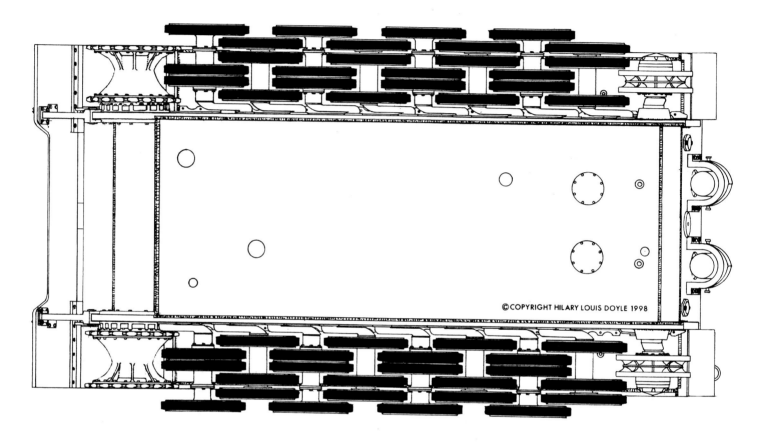

Panzerkampfwagen VI H Ausf.H1 – Fgst.Nr.V1 completed in April 1942 – underside of **Vorpanzer** – final drives without armor guard

OPPOSITE BELOW: The front of **V1** was the only **VK 45.01 (H)** completed with **Vorpanzer** (spaced armor). When fully opened, the cupola hatch lid rested on a flat stop welded to the side of the commander's cupola. (TTM)

Chapter 3: Panzerkampfwagen Tiger Ausf.E

The first **VK 45.01 (H)** (**Fgst.Nr.V1**) from the **Versuchsserie** in the assembly hall at Henschel in April 1942. **V1** had many unique features, including a horn mounted on the superstructure top plate, but did not have periscopes mounted in the driver's and radio operator's hatches. (TTM)

GERMANY'S TIGER TANKS - D.W. to TIGER I

The following sketches from a manual were used to identify components. They represent a Tiger I produced in mid-April 1943 after conversion to the Maybach HL 230 P45 engine.

Auspufftopf = Exhaust muffler
Einsteigluke = Crew hatch
Fahrersehklappe = Driver's visor
Kettenabdeckung = Track guards
Kraftstoffeinfuelloeffnung = Fuel filler cap
Kuehlluftaustritt = Cooling air outlet
Kuehllufteintritt = Cooling air intake
Kuehlwassereinfuelloeffnung = Radiator filler cap
Kuehlwasserueberdruckventil = Radiator safety valve
Kugelblende-MG = Machinegun ball mount
Motorraumbelueftung = Engine compartment ventilation
Pz-Fuehrerkuppel = Commander's cupola

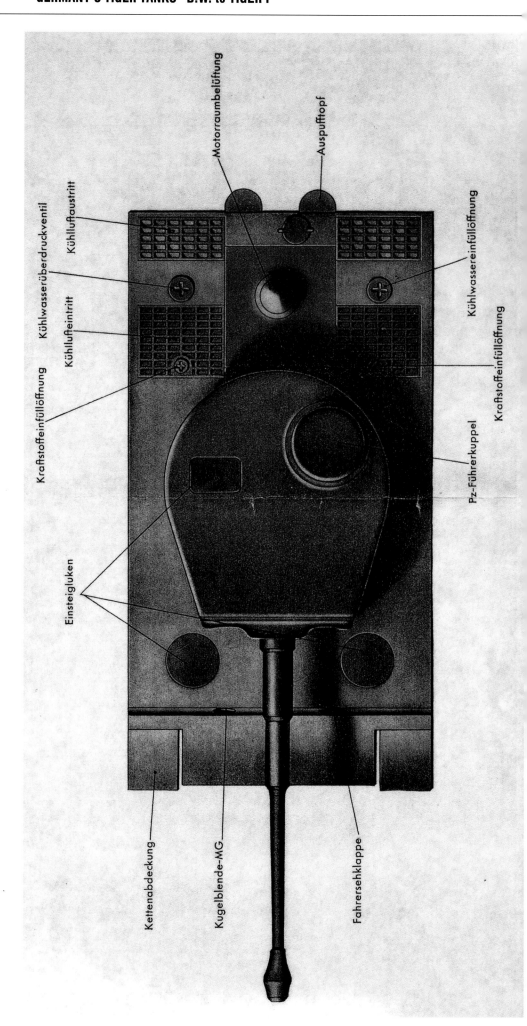

42

Chapter 3: Panzerkampfwagen Tiger Ausf.E

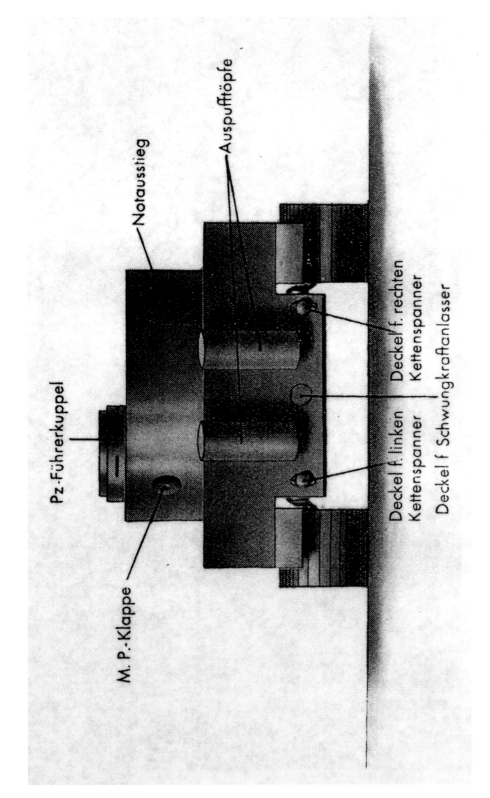

Auspufftoepfe = Exhaust mufflers
Deckel f. linken Kettenspanner = Cover for left track adjustment
Deckel f. rechten Kettenspanner = Cover for right track adjustment
Deckel f. Schwungkraftanlasser = Cover for inertia starter
M.P.-Klappe = Pistol port
Notaussteig = Escape hatch
Pz.-Fuehrerkuppel = Commander's cupola

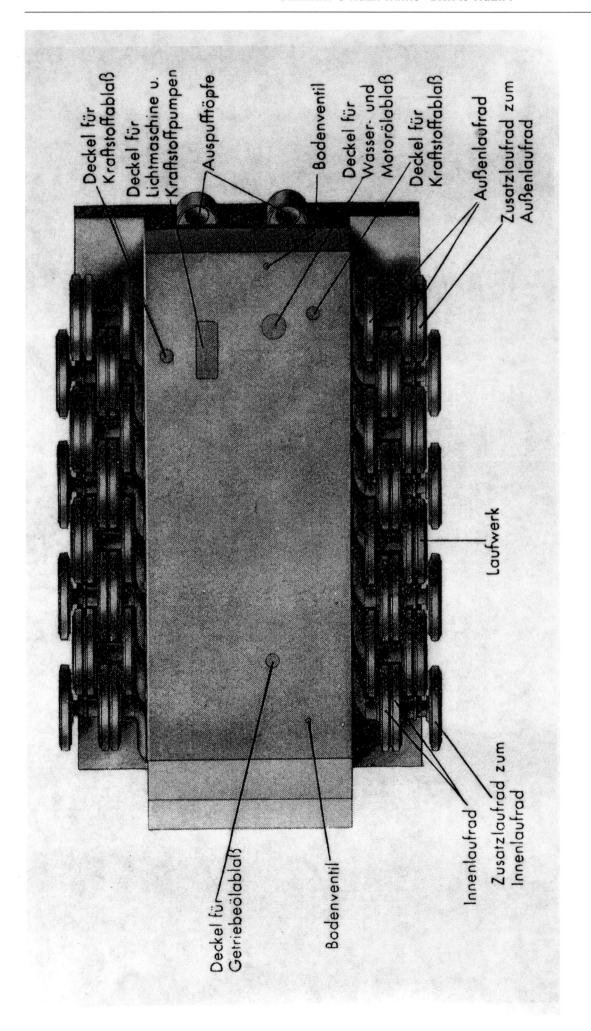

Auspufftoepfe = Exhaust mufflers
Aussenlaufrad = Outer roadwheel
Bodenventil = Drain valve
Deckel fuer Getriebeoelablass = Cover for draining transmission oil
Deckel fuer Kraftstoffablass = Cover for draining fuel
Deckel fuer Lichtmaschine und Kraftstoffpumpen = Cover for generator and fuel pumps
Deckel fuer Wasser- und Motoroelablass = Cover for draining water and engine oil
Innenlaufrad = Inner roadwheel
Zusatzlaufrad = Added roadwheel

Chapter 3: Panzerkampfwagen Tiger Ausf.E

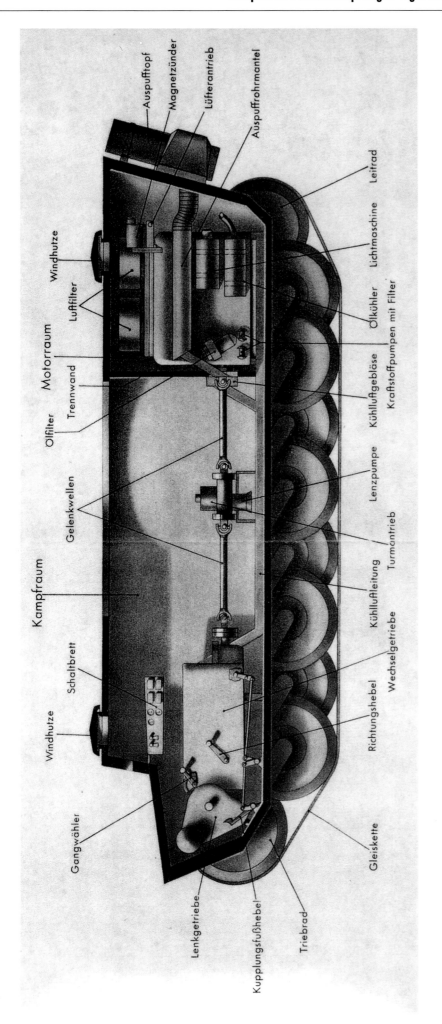

Auspuffrohrmantel = Exhaust pipe cover
Auspufftopf = Exhaust muffler
Gangwaehler = Gear selector
Gelenkwellen = Drive shafts
Gleiskette = Track
Kampfraum = Fighting compartment
Kraftstoffpumpe mit Filter = Fuel pump with filter
Kuehlluftgeblaese = Cooling air fan
Kuehlluftleitung = Cooling air duct
Kupplungsfusshebel = Foot pedal for clutch
Leitrad = Idler wheel
Lenkgetriebe = Steering gears
Lenzpumpe = Sump pump
Lichtmaschine = Generator
Luefterantrieb = Fan drive
Luftfilter = Air filter
Magnetzuender = Magneto
Motorraum = Engine compartment
Oelfilter = Oil filter
Oelkuehler = Oil cooler
Richtungshebel = Direction change lever
Schaltbrett = Instrument panel
Trennwand = Firewall
Triebrad = Drive wheel
Turmantrieb = Turret traverse drive
Wechselgetriebe = Transmission
Windhutze = Air intake armor guard

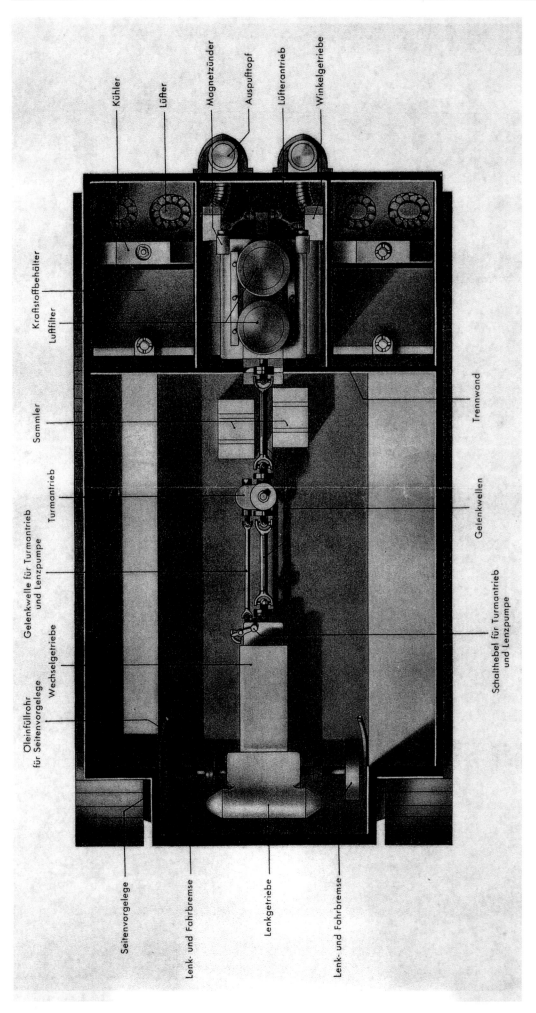

Auspufftopf = Exhaust muffler
Gelenkwellen = Drive shaft
Kraftstoffbehaelter = Fuel tank
Kuehler = Radiator
Lenkgetriebe = Steering gears
Lenk- und Fahrbremse = Steering and stopping brakes
Luefter = Fan
Luefterantrieb = Fan drive
Luftfilter = Air filter

Magnetzuender = Magneto
Oeleinfuellrohr = Oil filler pipe
Sammler = Battery
Schalthebel = Shift arm
Seitenvorgelege = Final drives
Trennwand = Firewall
Turmantrieb = Turret traverse drive
Wechselgetriebe = Transmission
Winkelgetriebe = Bevel gear drives

Chapter 3: Panzerkampfwagen Tiger Ausf.E

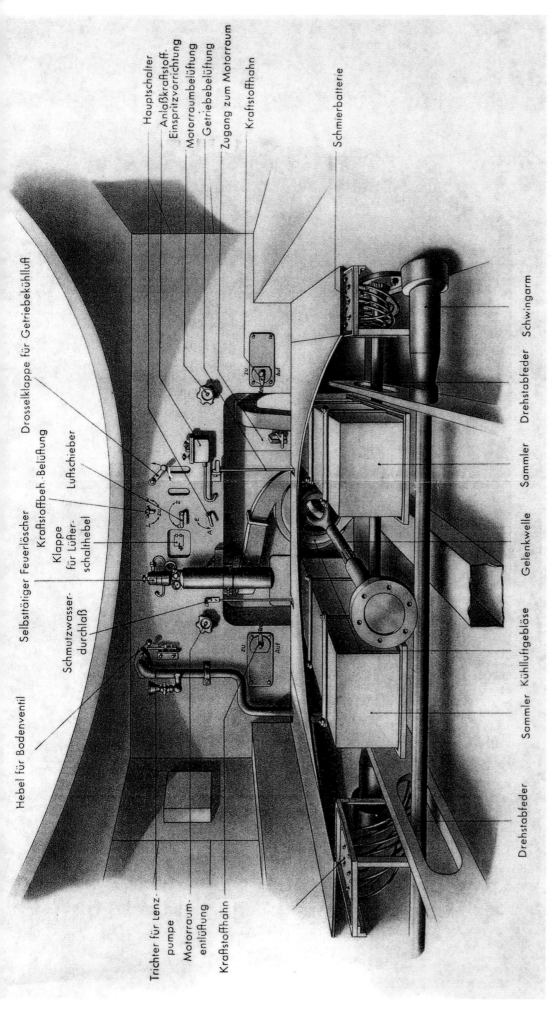

Anlasskraftstoff-Einspritzvorrichtung = Starter fuel injector
Drehstabfeder = Torsion bar
Drosselklappe fuer Getriebekuehlluft = Regulating flapper for transmission cooling air
Gelenkwelle = Drive shaft
Getriebebelueftung = Transmission ventilation
Hauptschalter = Master switch
Hebel fuer Bodenventil = Lever for drain valve
Klappe fuer Luefterschalthebel = Hatch to fan engaging lever
Kraftstoffbeh.-Belueftung = Fuel tank ventilation
Kraftstoffhahn = Fuel valve
Kuehlluftgeblaese = Cooling air fan
Luftschieber = Air damper
Motorraumbelueftung = Engine compartment air intake
Motorraumentlueftung = Engine compartment air outlet
Sammler = Battery
Schmierbatterie = Central lubrication
Schmutzwasserdurchlass = Dirty water drain
Schwingarm = Roadwheel arm
Selbsttaetiger Feuerloescher = Automatic fire extinguisher
Trichter fuer Lenzpumpe = Funnel for the sump pump
Zugang zum Motorraum = Access to engine compartment

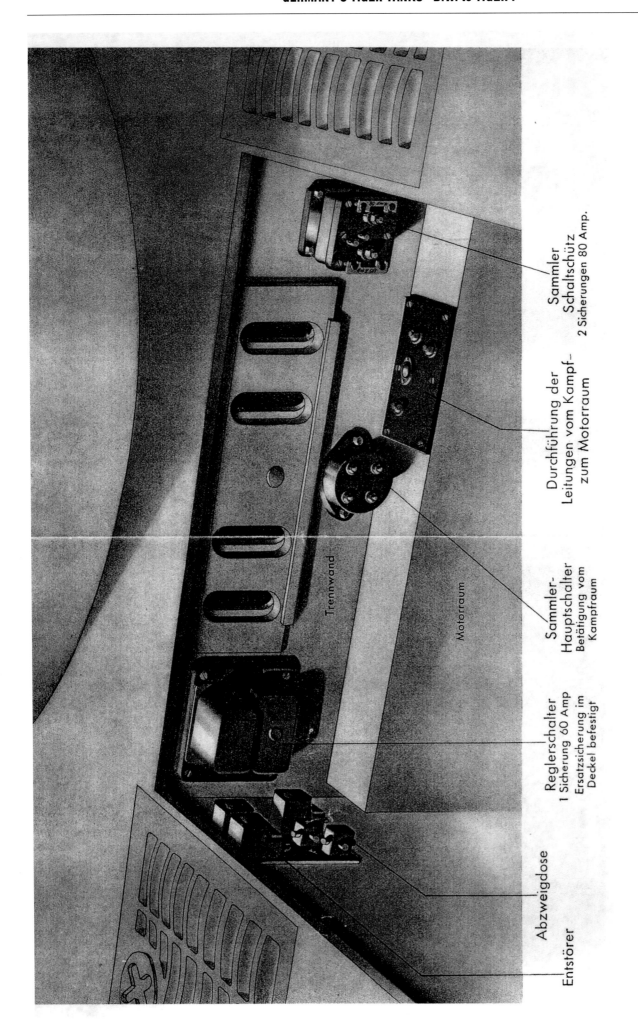

Abzweigdose = Distribution box
Betaetigung vom Kampfraum = Operated from crew compartment
Durchfuehrung der Leitungen = Penetrations
Entstoerer = Noise suppressor
Motorraum = Engine compartment
Reglerschalter = Regulator switch
Sammler-Hauptschalter = Master switch
Sammler-Schaltschuetz = Battery starting protection
Sicherungen = Fuses
Trennwand = Firewall

Chapter 3: Panzerkampfwagen Tiger Ausf.E

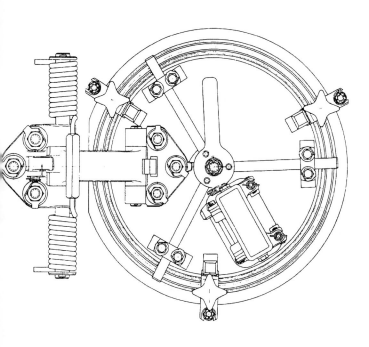

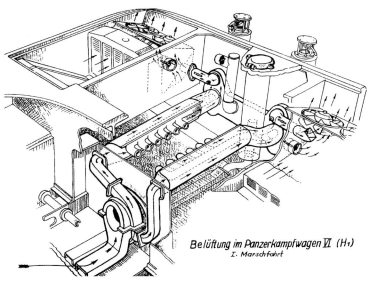

Ventilation flow paths during normal operation (above) and submerged fording (below) when the radiator and fan compartments are flooded.

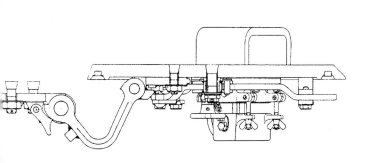

...ukendeckel (driver's and radio operator's hatches) with periscope and ...pring counterbalance.

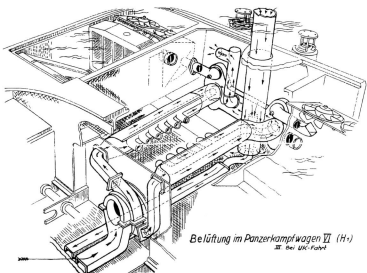

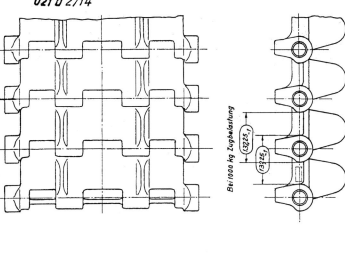

The first **HL 210 P45** (**Motor Nr.46051**) designed and produced by Maybach in 1942. It had three filters mounted on a manifold, and the magnetos were mounted at the rear over the valve covers.

LEFT: The 520 mm wide **Verladekette** (transport tracks) model **Kgs 63 520/130** for the **VK 45.01 (H)** was adopted from the **VK 36.01**.

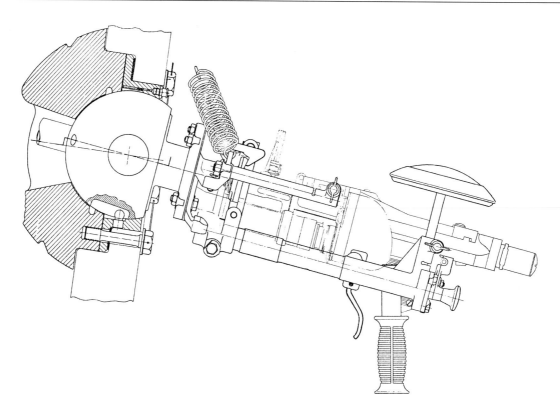

Kugelblende 100 (machinegun ball mount) with an **M.G.34 mit Panzermantel** (machinegun model 34 with armor barrel sleeve) and the **Kugelzielfernrohr 2** gunsight

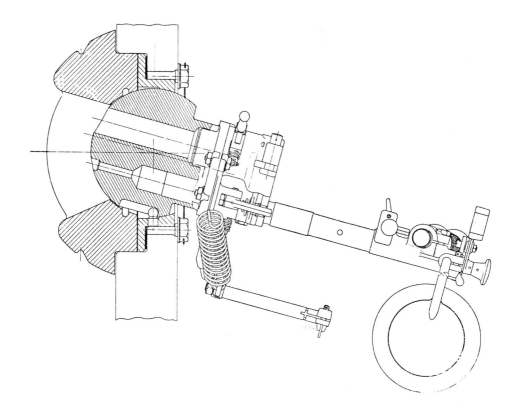

Chapter 3: Panzerkampfwagen Tiger Ausf.E

2.2 TURRET

The following description and drawings of the turret designed by Krupp were extracted from the manual **D 656/22, Panzerkampfwagen Tiger Ausf.E (Sd.Kfz.181), Panzerbefehlswagen Tiger Ausf.E, Geraetbeschreibung und Bedienungsanweisung zum Turm** dated 7 September 1944.

General Description – The turret, armed with an **8.8 cm Kw.K.36** and an **M.G.34**, is positioned in the middle of the Pz.Kpfw.Tiger. The gun tube, recoil cylinder, and pneumatic recuperator are mounted in the gun carriage. The **8.8 cm Kw.K.36** and **M.G.34** are mounted coaxially in the gun mantlet, which is elevated by a hand-operated elevation mechanism. The elevation arc extends from -8° to +15°. The forward weight of the gun tube and carriage is counterbalanced by a spring equilibrator. Spent casings ejected from the main gun are caught in a casing sack.

The turret can be traversed by the gunner using the hydraulic power traverse machine powered by an auxiliary drive or by hand. The turret is outfitted with a binocular **Turmzielfernrohr (T.Z.F.9b)** gun sight. With the aid of the 12-hour azimuth indicator system, the commander can communicate target selection commands to the gunner.

A number of **Pz.Kpfw.Tiger** are outfitted as **Panzerbefehlswagen**. For this purpose an additional radio set is installed in place of the coaxial machinegun.

The main components in the turret are:

Turm and Turmanschluss (turret and turret ring)
Walzenblende (gun mantlet)
Federausgleicher (spring counterbalance)
Geschuetzzurrung (internal travel lock)
MG-Lagerung (machinegun mount)
Turmzielfernrohr (gun sight)
Turmschwenkwerk (traverse drive)
Hoehenrichtmaschine (elevation mechanism)
Abfeuervorrichtungen (firing circuit)
Panzerfuehrerkuppel (commander's cupola)
Sehschlitzplatten (vision slits)
Turmlukendeckel (loader's hatch)
MP-Klappe (pistol port)
Turmsitze (crew seats)
Elektrische Einrichtung (electrical equipment)
Zubehoerlagerungen (equipment stowage)

2.2.1 Turm und Turmanschluss (Turret and Turret Ring)

The 100 mm thick turret front plates are angled at 85° from vertical, the 80 mm turret side plate at 0°, and the 25 mm roof plate at 85° and 90°. The leading edge of the turret roof is a thicker 40 mm plate extending the width of the turret behind the upper turret front plate. The turret side walls consist of a single plate bent into a horseshoe shape with the front closed by the upper and lower turret front plates. The rectangular opening left between the turret front and side plates is covered by the cast gun mantlet (varying from 85 to 200 mm thick). Refer to Appendix D for additional details on the armor specifications.

Sehschlitzplatten (vision slits) are inset into both left and right forward turret sides and **MP-Klappen** (pistol ports) both left and right toward the rear. The **Panzerfuehrerkuppel** is mounted to the left rear on the turret roof, the **Turmlukendeckel** (loader's hatch) to the right, and the **Aufbauluefter** (ventilation fan) to the rear. Three lifting studs are located on the turret sides for removing and mounting the turret.

A support ring, welded to the bottom of the turret, rests on the turret ball-bearing race. The 2100 mm outer diameter **Turmkugellager** (turret ball-bearing race) consists of two grooved steel rings, in between which run 79 supporting ball bearings (40 mm diameter) and 79 spacing ball bearings (39 mm diameter). The supporting ball bearings bear the turret weight and take up any side movement. The moving outer ball-bearing ring is fitted into and bolted to the turret support ring. The fixed inner ball-bearing ring is fitted into and bolted to the gear ring. The **Zahnkranz** (gear ring) with 240 teeth serves to turn the turret with the traverse drive. The gear ring is bolted to the armor hull and the inner ball bearing ring.

A **Dichtschlauch** (inflatable rubber inner tube) is located in the gap between the outer ball-bearing ring and the inner turret wall. When inflated with air, the rubber inner tube closes the gap watertight. Water seeping into the ball bearing race is drained out through a tube sealed by a threaded cap.

A sealing ring, mounted on the inside of the turret support ring, prevents dirt and sand from entering the ball bearing race. The inside of the gear ring is covered by a sheet metal guard.

The loader's seat, an elevation mechanism, and a bridge are bolted to a support on the right front turret ring. Under the gun, the bridge supports the handwheel and driveshaft for the elevation mechanism with the attached firing lever. To the left, the bridge supports the gunner's seat and commander's foot rest.

A **Zwoelfuhrzeiger** (azimuth indicator), marked from 1 to 12 o'clock, is mounted to the left front of the gunner on the turret ring and driven by the turret traverse mechanism.

The controls for the hydraulic drive for the turret traverse mechanism are mounted on the turret platform directly behind the sheet metal guard for the gunner's knees. The hydraulic turret drive is mounted on the turret platform and connected by a drive shaft to the traverse mechanism mounted on the turret ring.

The **Drehbuehne** (turret platform) is suspended from three supports fastened to the turret ring, turret traverse gear housing, and bridge support. The space underneath the turret platform is accessible through a hinged hatch in the platform floor.

3.2.2.2 Walzenblende (Gun Mantlet)

The gun cradle with the 8.8 cm gun tube, recoil cylinder, and recuperator, the **M.G.34** and the gun sight mount are all mounted in the gun mantlet. The gun mantlet, designed as a shield, has one penetration bored for the gun cradle, one for the **M.G.34** and two for the binocular **T.Z.F.9b** gun sight. Forward inside the protective armor tube, bolted to the face of the gun mantle, is a spring compressed seal which prevents dust from entering the gap around the 8.8 cm gun tube.

The gap between the turret front and the gun mantlet can be sealed watertight by installing a framed gasket. The gap between the gun tube and the gun cradle is made watertight with a cradle seal.

3.2.2.3 Federausgleicher (Spring Counterbalance)

The forward weight of the gun tube and carriage is counterbalanced by a spring equilibrator. Linked to the right side of the gun cradle, this spring equilibrator is mounted on the right side of the turret ring.

3.2.2.4 Geschuetzzurrung (Gun Travel Lock)

When not in use, the **8.8 cm Kw.K.36** is held stationary by an internal travel lock which is suspended under the turret roof. Two hooks on the travel lock hold studs on the sides of the breech block.

3.2.2.5 MG-Lagerung (Machinegun Mount)

The **M.G.34** mounting on the right side of the gun mantlet can be adjusted both horizontally and vertically. Two ammunition bags are suspended from a holder – one filled with 150 belted rounds and the second to catch ejected spent cartridges. The opening for the **M.G.34** in the gun mantlet can be sealed watertight with a plug after the **M.G.34** is dismounted.

3.2.2.6 Turmzielfernrohr 9b (Binocular Gun Sight)

The **T.Z.F.9b** is a jointed, binocular gun sight for direct fire, with an armor plate at the joint to protect the gunner. In contrast to a simple monocular sight, binocular sights made it possible to observe and aim for longer periods without tiring.

The sighting angle for the various ranges is changed by adjusting the sighting mark seen in the viewing field. Each of the two telescopes has an optical length of 814 mm. The magnification is 2.5x with a field of view of 25° (equal to 444 meters wide at a range of 1000 meters) through the 5 mm diameter exit pupil. Focus is adjustable to the gunner's vision.

The reticle in the right-hand telescope consists of a central inverted V mark flanked by three smaller inverted Vs to the right and left spaced at 4 mil intervals to aid in leading moving targets. Range markers are inscribed in an arc for the various types of ammunition up to 4000 meters. A reticle (with only the inverted V marks) in the left-hand telescope can be rotated into place if the right-hand telescope is unusable. An adjustable light is provided to illuminate the reticles in dim light.

The front of the binocular **T.Z.F.9b** is mounted on the slides in the gun sight mount, and the rear is suspended from the turret roof by a pivoting arm. The forward gun sight mount is fastened to the gun carriage and pivoted with it. For underwater travel, both of the openings in the turret front plate for the binocular gun sight can be sealed with plugs.

3.2.2.7 Turmschwenkwerk (Traverse Drive)

The turret can be traversed by hydraulic power or turned by hand. A power takeoff from the transmission drives the hydraulic motor for the traverse drive. Powered traverse is controlled by a foot pedal mounted on the turret platform. A hand-operated clutch is used to engage and disengage the drive. The fastest that the turret can traverse by hydraulic power is one full circle of 360° in 60 seconds. One full turn of the gunner's hand wheel results in traversing the turret 0.5° or about 9 mils.

3.2.2.8 Hoehenrichtmaschine (Elevation Mechanism)

The elevation mechanism moves the gun mantlet as the gunner turns a handwheel with his right hand. One complete turn of the hand wheel changes the gun's elevation by about 0.97° or 1 mils. The gun trigger and firing circuit switch are located close behind the elevating hand wheel.

3.2.2.9 Abfeuervorrichtungen (Firing Circuit)

A firing lever located close behind the elevating hand wheel allows the gunner to fire the main gun without releasing the hand wheel. The coax **M.G.34** is fired by the gunner depressing a foot pedal that is connected by a complex linkage to a pivotal trigger bar.

3.2.2.10 Pz-Fuehrerkuppel (Commander's Cupola)

The **Pz-Fuehrerkuppel** is bolted to the left rear of the turret roof. It serves as the entrance hatch and provides sheltered observation for the commander. The **Pz-Fuehrerkuppel** consists of a cylindrical mantle with five vision block holders, the azimuth indicator ring, and the hatch lid. Five vision slits are cut into the cylindrical mantle. Behind each vision slit is a 90 mm thick laminated glass block to protect the eyes against bullet splash and small arms fire. The frame and locking lever holding the vision blocks in place are designed so that the vision blocks can be easily replaced. Head and nose guards are mounted on each frame.

A single wire welded to the outside of the forward vision slit serves as the front sight, and twin wires on the frame for the vision block serve as a rear sight. With this sighting arrangement and the azimuth indicator ring, the commander can tell the direction in which the turret weapons are aimed and direct the gunner onto the target. The azimuth indicator ring, numbered from 1 through 12, is mounted on three ball bearings and turned by a drive shaft connected to the turret ring gears.

The hatch lid for the **Pz-Fuehrerkuppel** is fastened to the mounting ring by double hinges. Three latches on the underside of the hatch lid are used to secure the hatch lid to the mounting ring. A lock, which can be closed and opened with a square key from the outside, is used to close the hatch lid when the crew leave the Panzer. A padlock, inserted through the hasp welded to the mounting ring, can also be used to lock the hatch closed. A handle for opening the hatch lid is welded to the outside.

3.2.2.11 Sehschlitzplatten (Vision Slit)

Vision slits are built into the forward turret sides on both the left and right sides. The **Sehschlitzplatten** are welded to the turret side. A 90 mm thick laminated glass block is mounted in a holder behind each vision slit. The vision block is made watertight by being pressed into a rubber gasket by retaining screws on each side of the hinged frame. Head and nose guards are mounted on each frame.

3.2.2.12 Ladeschuetzenlukendeckel (Loader's Hatch)

The hatch in the turret roof is used for entering and leaving the turret. The hatch cover, pivoting on two hinges, is mounted on a frame bolted to the turret lid. There is one handle on the outside and two on the inside. To close the hatch watertight, four locking bars are pushed outward by a central ring and tightened by the central threaded spindle pulling the hatch cover down onto the rubber gasket.

Chapter 3: Panzerkampfwagen Tiger Ausf.E

A forked latch can be used to lock the hatch lid closed from the inside or outside using a square key. A padlock, inserted through the hasp welded to the outside, can also be used to lock the hatch closed.

3.2.2.13 MP-Klappe (MP Port)

Two **MP-Klappe** are built into the turret side wall – one on the left rear and one on the right rear. The port in the **MP-Klappe** is designed to be opened and closed by rotating an internally mounted armor shutter. A plug is inserted into the opened **MP-Klappe** to seal it watertight before submerged fording.

3.2.2.14 Turmsitze (Crew Seats)

Pz-Fuehrersitz (Commander's Seat) – Two seats with cushions are mounted in the left turret rear on a stanchion secured to the turret ring. The upper seat, used by the commander to look out of the open cupola, can be folded down and used as a backrest when the commander sits in the lower seat to look though the cupola vision blocks.

Richtschuetzensitz (Gunner's Seat) – A fixed seat with a cushion is bolted to the left end of the bridge in the turret. The back rest is mounted on a pole which is clamped to the bridge.

Ladeschuetzensitz (Loader's Seat) – The backrest and cushioned seat are attached to a pole which is bolted to the block supporting the elevation mechanism on the right side of the turret. When not in use, the seat can be pivoted out of the loader's way.

3.2.2.15 Elektrische Einrichtung (Electric Equipment)

The firing circuit is powered by the 12 volt vehicle battery. Originally there was no emergency firing device for use when the vehicle power failed.

Turret lighting is provided by three lights mounted under the roof. The amount of light can be adjusted by rotating covers which turn off the light when they are completely closed. Lights are also provided for the gun sight reticles and the azimuth indicator. A receptacle is available to plug in a hand-held lamp.

A ventilator fan with a flow rate of 12 m^3/min is mounted under the turret roof. It served to suck out any fumes left in the turret after the guns were fired.

3.2.3 EQUIPMENT

The following list of equipment was extracted from the *Vorlaeufige Ausruestung fuer Aufbau des Pz.Kpfw.VI H, Ausfuehrung H1* from 1942:

1 **T.Z.F.9b** (binocular telescopic gun sight)
1 **K.Z.F.2** (gun sight for machinegun ball mount)
1 **K.F.F.2** (driver's periscopes)
3 70x270x94 protective glass blocks
 (1 for driver's visor, 2 spare)
11 70x150x94 protective glass blocks
 (2 for turret vision slits, 5 for cupola vision slits, and 4 spare)
2 **M.G.34 mit Panzermantle** (machineguns with armor sleeve)
 (1 in turret and 1 in hull)
32 **Gurtsack** (bags for machinegun ammunition belts)
 (22 in **Befehlswagen**)
1 **M.P.** (machinepistol)
1 **Leuchtpistole** (signal pistol)
24 **Leuchtpatronen** (signal flares) (12 white, 6 red, 6 green)
1 **Kurskreisel** (gyroscopic compass)
1 **Ausfallflagge (gelb-schwarz)** (out-of-action flag)

3.2.4 COMMUNICATIONS

When assigned to company headquarters or to a platoon leader, the **Panzerkampfwagen Tiger (8.8 cm Kw.K.36 L/56) (Sd.Kfz.181)** was outfitted with two radio sets, a **Fu 5** (10 watt transmitter with ultra short wave length receiver, operated in the frequency band 27.2 to 33.4 MHz) and a **Fu 2** (ultra short wave length receiver, operating in the same frequency band as the **Fu 5**). The **Fu 5** had a usable range of 4 to 6 kilometers, highly dependent on terrain and atmospheric conditions. The other nine **Panzerkampfwagen Tiger (8.8 cm Kw.K.36 L/56) (Sd.Kfz.181)** in each company of 14 were outfitted with a single **Fu 5** radio set. A **Bordsprechanlage** (intercom system) was to be installed in all **Panzerkampfwagen Tiger (8.8 cm Kw.K.36 L/56) (Sd.Kfz.181)**. Only four of the crew members had speaker headsets and a microphone. As shown in the wiring diagram for the turret, provisions had not been made for a speaker headset or microphone for the loader. The commander could also use voice, signal pistol, flags, orange smoke signals, and directional fire for communicating with others outside the Tiger I.

3.2.5 PANZERBEFEHLSWAGEN

A number of **Pz.Kpfw.Tiger** are outfitted as **Panzerbefehlswagen** with additional radio equipment mounted in the turret. The space needed for mounting and operating the radio equipment is acquired by dropping the machinegun mount. A **Pz-Funkersitz** (radio operator seat) is installed in place of the loader's seat. The vision slit in the right turret side, the opening for the machinegun in the gun mantlet, and (later) the hole in the turret roof for the loader's periscope are sealed with armor plugs welded in place. The **Panzerbefehlswagen Tiger** is outfitted with two different radio sets: **Sd.Kfz.267** with **Fu 5** and **Fu 8** or **Sd.Kfz.268** with **Fu 5** and **Fu 7**.

When the **Tiger** is outfitted as a **Panzerbefehlswagen**, the following equipment is dropped: 1 **M.G.34** in the gun mantlet, 1 spare barrel holder, 1 box for machinegun tools, 1 box for machinegun accessories (sight and bipod), 10 belt bags for machinegun ammunition and the stowage brackets (1500 rounds of machinegun ammunition), 26 rounds of 8.8 cm ammunition and the stowage racks, 1 baggage box in the right front of the turret by the loader, and (later) 1 periscope for the loader.

The following radio equipment is installed: **Bordsprechanlage B** (intecom set model B) for **Panzerbefehlswagen**, 1 radio set **Fu 5 (10 W-Sender c und UKW-Empfaenger e)** in the turret, 1 radio set **Fu 7 (20 W-Sender d und UKW-Empfaenger d1)** or **Fu 8 (30 W-Sender a und MW-Empfaenger c)** in the hull, 1 **Maschinensatz GG 400** (electrical generator set), 1 radio accessories box, 1 antenna connector for the **Sternantenne D** for the **Fu 8** (mounted on the right side on the rear deck), 1 **Stabantenne 1.4 m** for the **Fu 7** (mounted on the left side on the rear deck), 1 **Stabantenne 2 m** for the **Fu 5** (mounted on the turret roof), and (later) a tube for stowing the **Steckmastrohre** (antenna extension rods) and spare antennas (mounted outside on the hull rear). All three antennas are mounted on **Gummiantennenfuessen** that can be bent in all directions and spring back up by themselves. The rubber-cushioned radio rack frames for the **Fu 7** or **Fu 8** sender and receiver radio sets are mounted one above the other to the left of the radio operator in the hull. The rubber-cushioned radio rack frames for the **Fu 5** sender and receiver radio sets are mounted one above the other on the right turret wall.

Panzerbefehlswagen armament consists of 1 **8.8 cm Kw.K. L/56** in the gun mantlet, 1 **M.G.34** in the ball mount in the driver's front plate, (later) 1 **M.G.34** as a **Flieger-MG** (anti-aircraft machinegun) in the turret, and 1 **MP** (machine pistol). Ammunition stowage consists of 66 rounds for the **8.8 cm Kw.K.**, 22 belt bags each with 150 rounds of machinegun ammunition, and 1 signal pistol flares.

The **Panzerbefehlswagen** is manned by a crew of five consisting of a **Kommandeur (Panzerfuehrer)** (commander), **Nachrichtenoffizier (Richtschuetze)** (signals officer – gunner), **Panzerfunker 1 (Ladeschuetze)** (radio operator – loader), **Panzerfunker 2 (Panzerfunker)** (radio operator) and **Fahrer** (driver).

Chapter 3: Panzerkampfwagen Tiger Ausf.E

The following altered photographs and sketches from a manual were used to identify components. They represent a Tiger I produced in mid-April 1943 after adding the loader's periscope and spare track hangers.

8,8 cm KwK 36 = 88 mm main gun
Drehbuehne = Turret platform
Fluessigkeitsgetriebe = Hydraulic drive
Fusshebel fuer MG-Abzug = Foot pedal to fire machinegun
Ladeschuetzensitz abgeklappt = Raised loader's seat
Notausstiegklappe geoeffnet = Open escape hatch
Pz.-Fuehrerkuppel = Commander's cupola
Richtschuetzensitz = Gunner's seat
Schutzglaesern = Vision blocks
Stuetzwinkel fuer Gepaekkasten = Support for baggage bin
Tragzapfen = Lifting lug

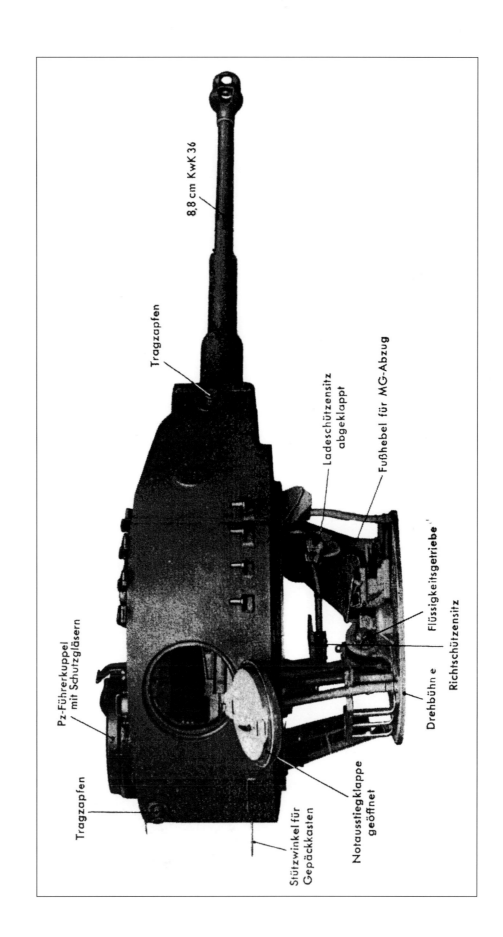

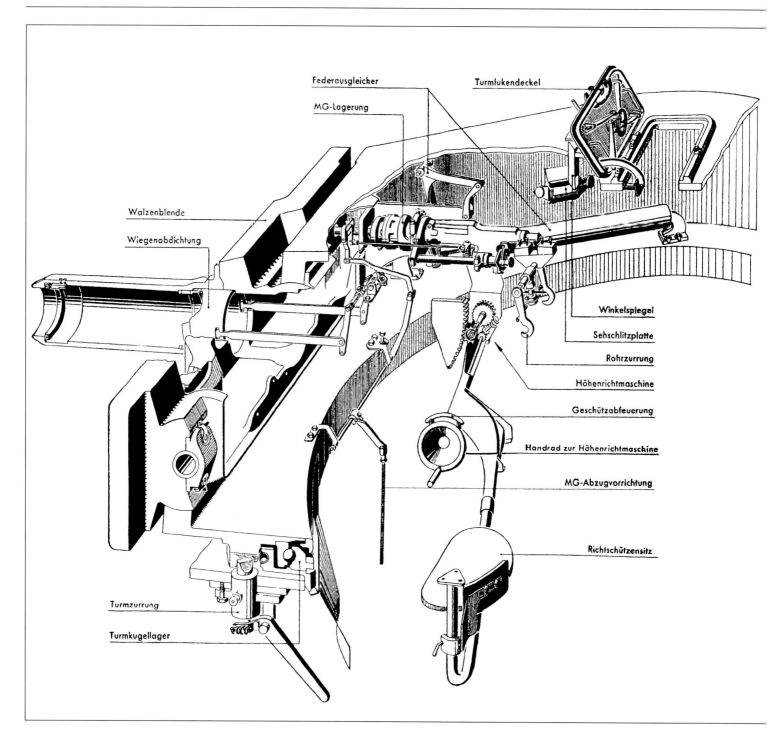

Federausgleicher = Spring counterbalance
Geschuetzabfeuerung = Trigger
Handrad = Hand wheel
Hoehenrichtmaschine = Elevating gear
MG-Abzugvorrichtung = Machinegun firing linkage
MG-Lagerung = Machinegun mount
Richtschuetzensitz = Gunner's seat
Rohrzurrung = Gun travel lock
Sehschlitzplatte = Vision slit plate
Turmkugellager = Ball-bearing race
Turmlukendeckel = Loader's hatch lid
Turmzurrung = Turret travel lock
Walzenblende = Gun mantlet
Wiegenabdichtung = Carriage seal for submerged fording
Winkelspiegel = Periscope

Chapter 3: Panzerkampfwagen Tiger Ausf.E

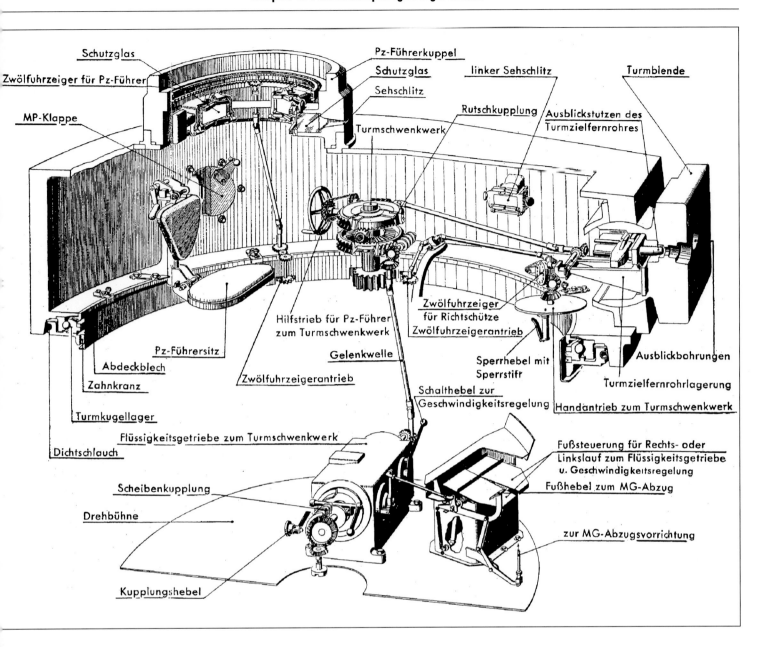

Abdeckblech = Cover guard
Ausblickbohrungen = Gunsight apertures
Ausblickstutzen = Front end support
Drehbuehne = Turret platform
Dichtschlauch = Sealing inner tube
Fluessigkeitsgetriebe = Hydraulic drive
Fusshebel zum MG-Abzug = Foot pedal to fire machinegun
Fussteuerung = Right and left steering foot pedal
Gelenkwelle = Drive shaft
Handantrieb = Hand drive for turret traverse
Hilfstrieb = Auxiliary hand traverse
Kupplungshebel = Clutch disengaging rod
Linker Sehschlitz = Left vision slit
MG-Abzugsvorrichtung = Machinegun firing linkage
MP-Klappe = Pistol port
Pz.-Fuehrerkuppel = Commander's cupola
Pz.-Fuehrersitz = Commander's seat
Rutschkupplung = Slip clutch
Schalthebel zur Geschwindigkeitsregelung = Traverse speed selector
Scheibenkupplung = Multi-plate clutch
Schutzglas = Vision block
Sehschlitz = Vision slit
Sperrhebel mit Sperrstift = Locking pin release handle
Turmblende = Gun mantlet
Turmkugellager = Ball bearing race
Turmschwenkwerk = Turret traverse drive
Turmzielfernrohres = Binocular gun sight
Turmzielfernrohrlagerung = Gun sight mount
Zahnkranz = Gear ring
Zwoelfuhrzeiger = 12-hour azimuth indicator
Zwoelfuhrzeigerantrieb = Azimuth indicator drive

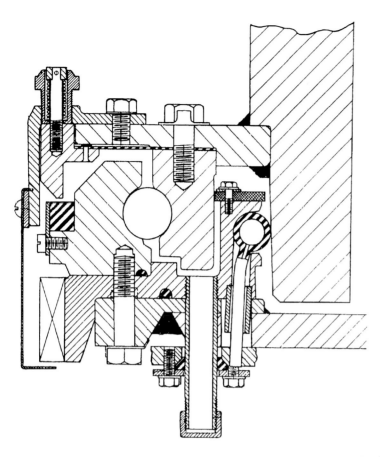

Turmkugellager (turret ball-bearing race) consisting of grooved steel rings in between which run 79 supporting ball bearings and 79 spacing ball bearings. A **Dichtschlauch** (inflatable rubber inner tube) was located in the gap between the outer ball-bearing ring and the inner turret wall.

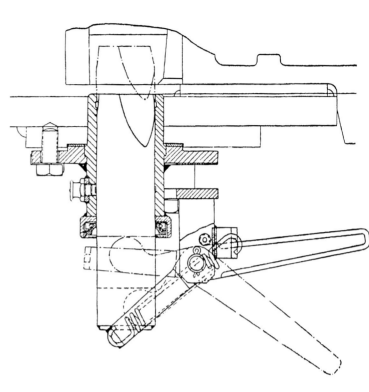

The **Turmzurrung** (turret traverse lock) caused problems when it vibrated out, allowing the gun to swing around and strike the driver's or radio operator's open hatches.

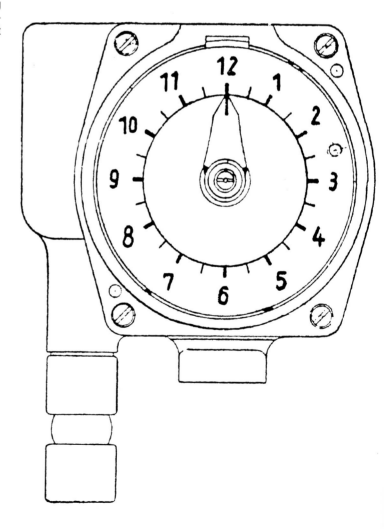

Zwoelfuhrzeiger (azimuth indicator), marked from 1 to 12 o'clock, was mounted to the left front of the gunner on the turret ring.

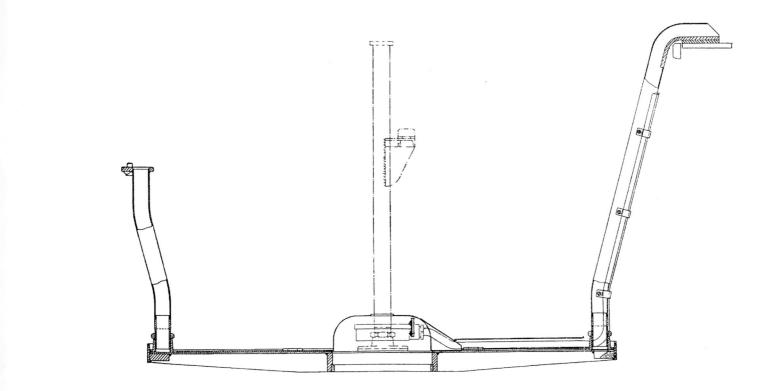

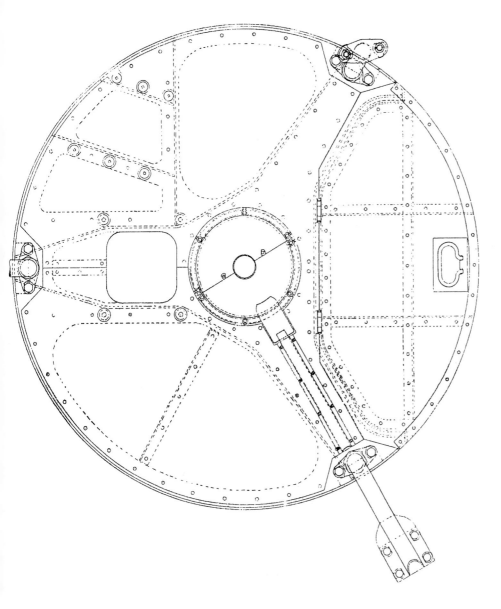

The **Drehbuehne** (turret platform) suspended from three supports fastened to the turret ring, turret traverse gear housing, and bridge support.

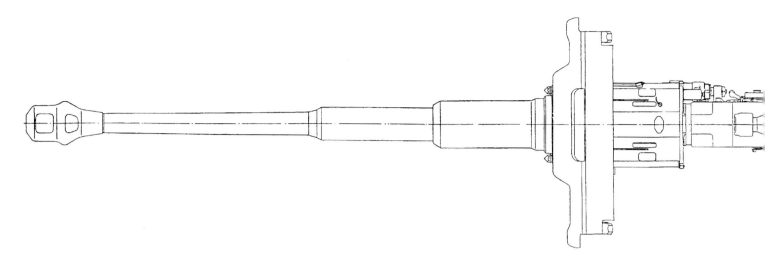

The **8.8 cm Kw.K.36 L/56** mounted in the **Walzenblende** (gun mantlet) with the muzzle brake drawn as a side view.

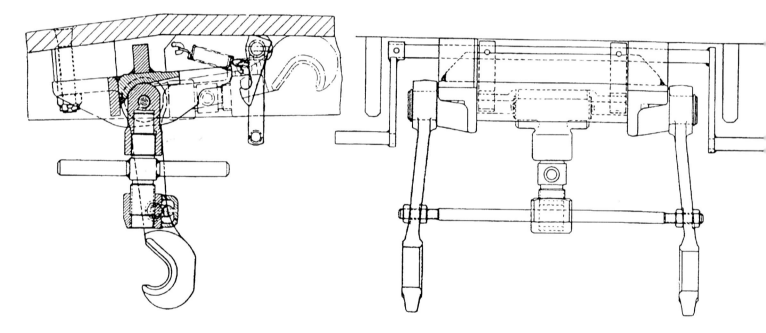

When not in use, the **8.8 cm Kw.K.36** was held stationary by the **Geschuetzzurrung** (internal gun travel lock), which was suspended under the turret roof.

Chapter 3: Panzerkampfwagen Tiger Ausf.E

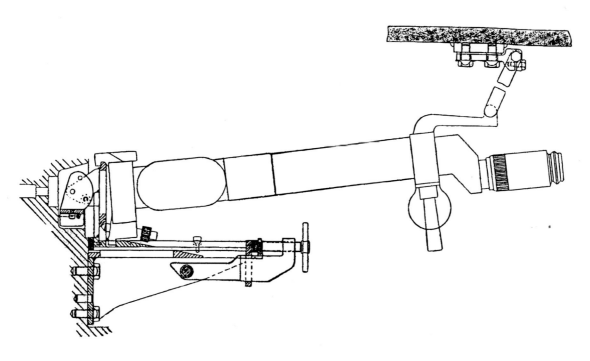

The **Turmzielfernrohr 9b** (binocular gunsight) was articulated so that the rear remained stationary as the front pivoted with the gun.

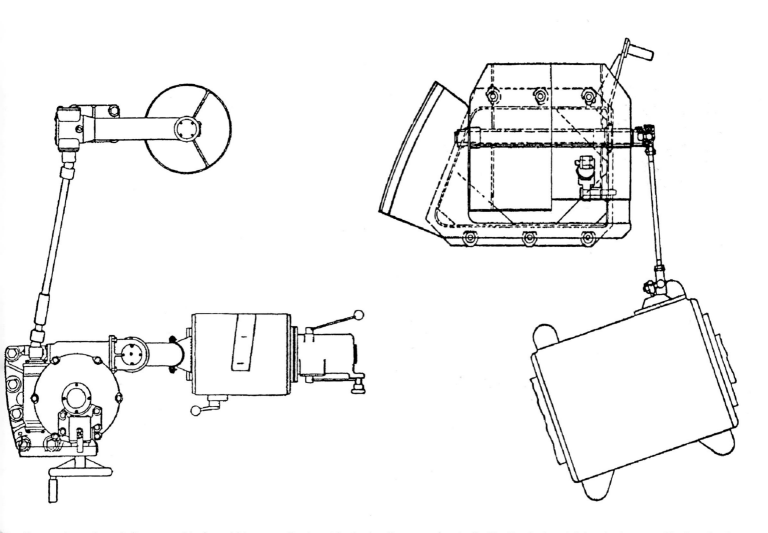

The **Turmschwenkwerk** (traverse drive) could traverse the turret by hydraulic power (controlled by the foot pedals) or be traversed by hand using the hand wheels.

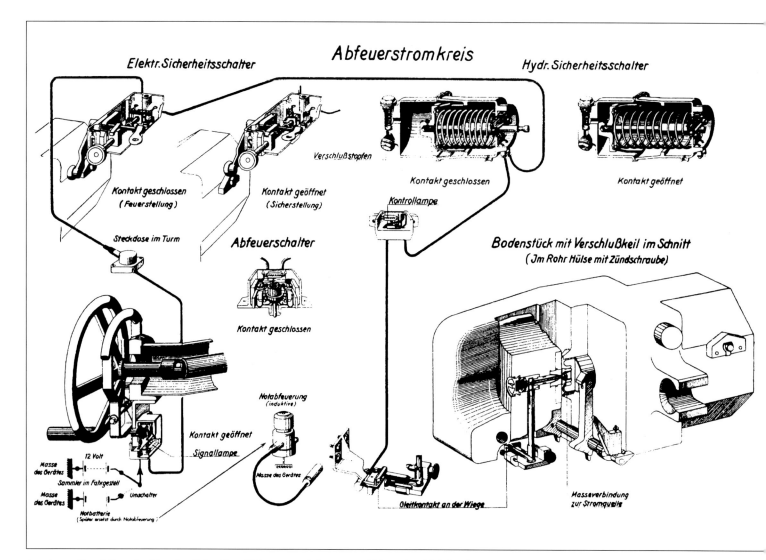

A schematic of the electric circuit for the **Abfeuervorrichtungen** (firing circuit). The firing lever located close behind the elevation hand wheel allowed the gunner to fire the main gun without releasing the elevation hand wheel.

Chapter 3: Panzerkampfwagen Tiger Ausf.E

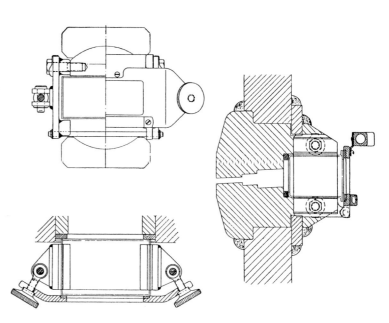

The **Sehschlitzplatten** (vision slit plates) were welded to the turret side. A 90 mm thick laminated glass block was mounted in the frame behind each vision slit.

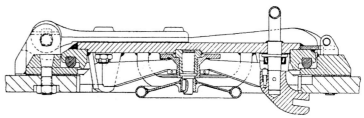

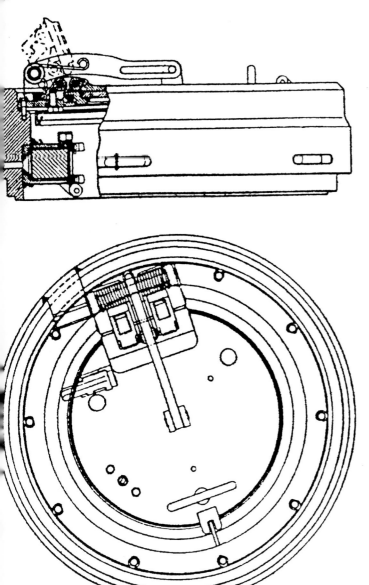

ABOVE AND BELOW: The **Pz-Fuehrerkuppel** (commander's cupola) consisted of a cylindrical mantle with five vision block holders, the azimuth indicator ring, and the hatch lid. The cupola was bolted to a ring welded to the turret roof. This drawing shows the cupola after it was modified to add an arm to secure the opened lid at 110 degrees and a **Federausgleicher** (Spring Counterbalance).

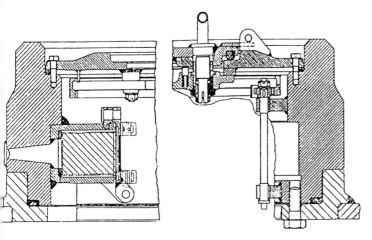

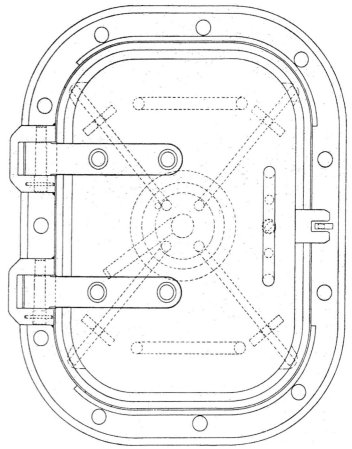

RIGHT: The **Ladeschuetzenlukendeckel** (loader's hatch lid) pivoted on two hinges. When closed it was protected by a bullet deflector frame that was bolted to the turret roof. The "forked" latch could be used to leave the hatch cracked open for the loader to get fresh air.

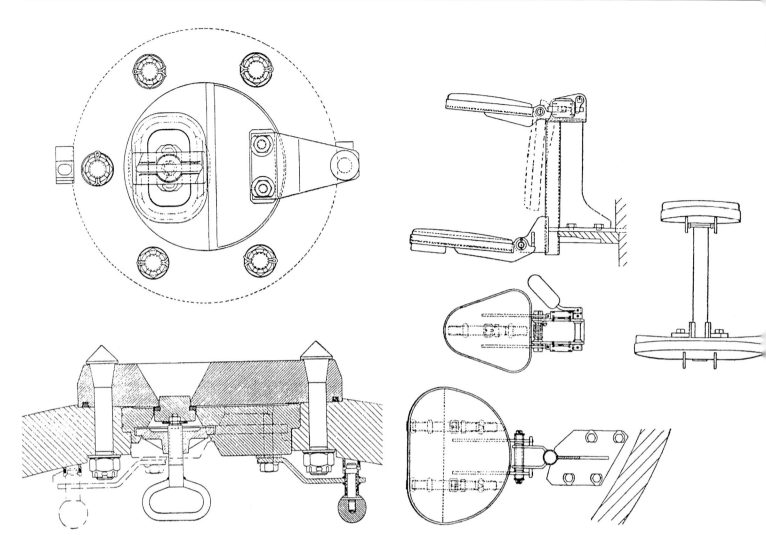

A plug was inserted into the opened **MP-Klappe** (pistol port) to seal it watertight before submerged fording.

The **Pz-Fuehrersitz** (commander's seat) consisted of two seats, an upper to look out the open cupola and a lower to look though the cupola vision blocks.

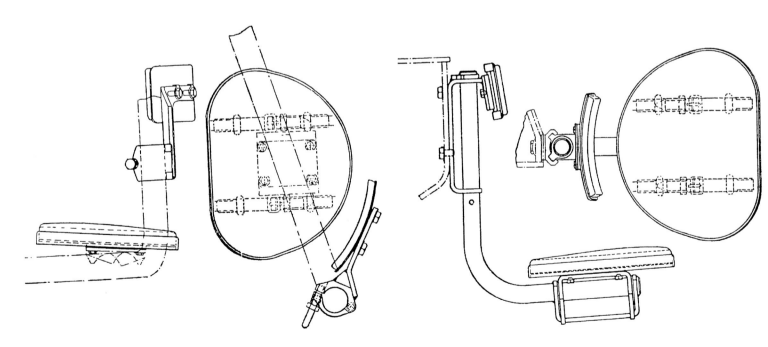

The **Richtschuetzensitz** (gunner's seat) was a fixed seat mounted on the bridge with a side back.

The **Ladeschuetzensitz** (loader's seat) could be pivoted out of the way when not in use.

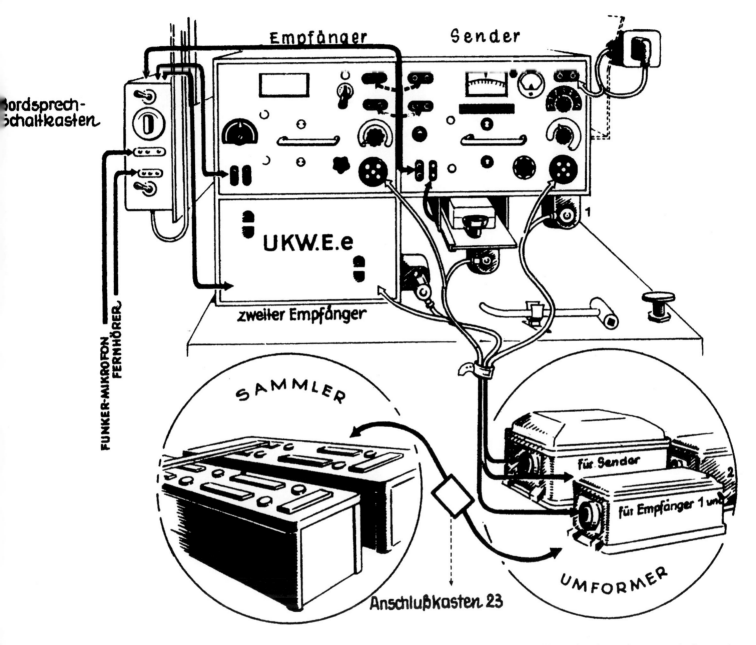

Both an **Fu 5** (10 watt transmitter and receiver) and an **Fu 2** (ultra short wave length receiver) were mounted in racks above the transmission to the left of the radio operator.

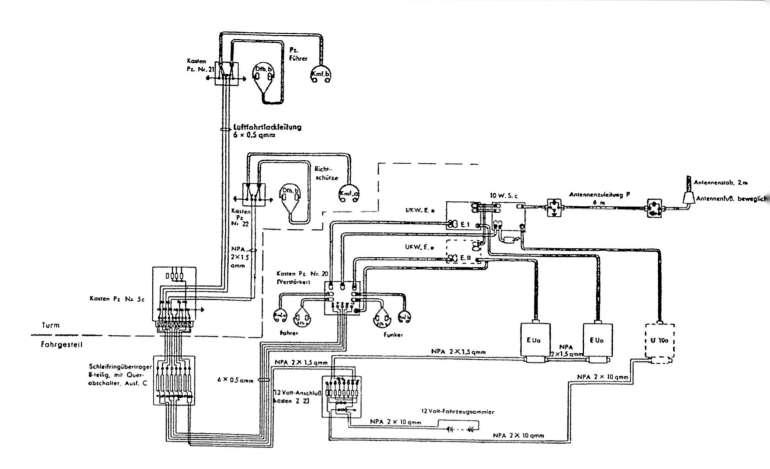

The wiring diagram for the radio sets in a normal **Pz.Kpfw.Tiger** assigned to company headquarters or to a platoon leader (the other Tigers had only one sender set and one receiver set). The commander, gunner, driver, and radio operator (but not the loader) had speaker headsets and throat microphones.

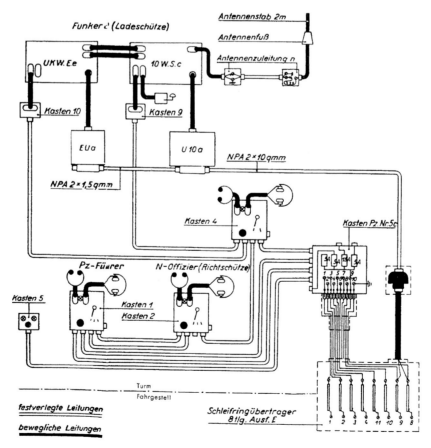

The wiring diagram for the **Fu 5** radio set mounted in the turret of a **Panzerbefehlswagen Tiger**. All three crew members in the turret – **Pz-Fuehrer** (commander), **N-Offizier** (gunner), and **Funker 2** (loader) had headsets and microphones.

Chapter 3: Panzerkampfwagen Tiger Ausf.E

3 TIGER I PRODUCTION

3.1 FINAL ASSEMBLY

Henschel was awarded contract number SS 006-6307/41 by Wa Pruef 6 in July 1941 to assemble three **Versuchs-Fahrgestell** (trial chassis) with **Fgst.Nr. V1, V2,** and **V3**. Henschel's first series production contract (number SS 4911-210-5904/41) was awarded to them by **Wa J Rue (WuG 6)** for assembly of 100 **VK 45.01(H)** chassis with **Fgst.Nr.250001-250100**. Before the first chassis had been completed in April 1942, **Wa J Rue (WuG 6)** issued another contract (SS 4911-210-5910/41) for Henschel to assemble 200 additional **VK 45.01(H)** chassis. These 200 were to be produced as **Pz.Kpfw.VI H Ausf.H2** with a **7.5 cm Kw.K.42 L/70** gun in turrets designed by Rheinmetall. On 14 July 1942, the decision was made not to outfit the **VK 45.01(H)** with **7.5 cm Kw.K.42 L/70** guns, therefore this second and all subsequent production series still had turrets mounting the **8.8 cm Kw.K. L/56** gun.

An additional 124 **VK 45.01(H)** were ordered in August 1942, bringing the total for the production series up to 424. In October 1942, the decision was made that Henschel would start production of the **VK 45.03** in September 1943. Thereafter, 250 **VK 45.01(H)** were ordered in November 1942, 490 by March, and 128 by 12 April 1943 to keep the Henschel assembly plant working at maximum capacity until **VK 45.03** production could ramp up. The last series production order had been expanded to 128 by February 1944. The final batch of **Tiger I** (originally 45, then increased to 54) were produced mainly by recycling armor components from Tigers that had been too extensively damaged for repairs at the front and returned to Germany for major overhaul.

Wegmann Waggonfabrik A.G. in Kassel was awarded contracts to assemble turrets and deliver them to Henschel at the same time that Henschel was awarded the following contracts for assembling the chassis in automotive running order:

Contract Number	Number Ordered	Chassis Number
SS 006-6307/41	3	V1 - V3
SS 4911-210-5904/41	100	250001 - 250100
SS 4911-210-5910/41	200	250101 - 250300
SS 4911-210-5910/41	124	250301 - 250424
SS 4911-210-5910/41	250	250425 - 250674
SS 4911-210-5910/42	490	250675 - 251164
SS 4911-210-5910/42	128	251165 - 251292
SS 4911-210-5910/42	54	251293 - 251346

Refer to Section 6.2.1 for a description of the Henschel & Sohn Werk III assembly plant in Kassel-Mittelfeld and Section 6.2.2 for a description of the Wegmann Waggonfabrik A.G. assembly plant in Kassel.

Krupp sent **Wanne Nr.1** (armor hull number 1) to Henschel on 3 January 1942 and **Turm Nr.1** (turret number 1) complete with **8.8 cm Kw.K.36 Rohr Nr.1** and **T.Z.F.9b Nr.3** to Henschel

Pz.Kpfw.VI H Ausf.H1 hulls on the assembly line at Henschel on 5 September 1942. (TTM)

Tiger I Production

Month	Week 1	Week 2	Week 3	Week 4	Week 5	Total	Goal for Month	Accepted by WaA	Chassis No.
	\multicolumn{5}{c}{Reported by Henchel}				
April 1942						1	1	0	V1
May 1942						0	3	0	
June 1942						0	5	0	250001
July 1942						0	10	0	
August 1942						8	10	8	250009
September 1942						3	15	3	250012
October 1942						11	16	10	250021+V2
November 1942						25	18	17	250038
December 1942						30	30	38	250076
January 1943	5+1R	10	8	7	5	35	30	35	250111+V3
February 1943	0	9	9	10	4	32	30	32	250143
March 1943	4	5	8	13	11	41	40	41	250184
April 1943	0	7	15	10	14	46	45	46	250230
May 1943	0	8	12	4	26	50	50	50	250280
June 1943	4	12	9	13	22	60	60	60	250340
July 1943	0	7	15	17	26	65	65	65	250405
August 1943	0	3	16	17	24	60	70	60	250465
September 1943	8	15	13	27	22	85	75	85	250550
October 1943	0	10	18	14	8	50	80	50	250600
November 1943	4	0	22	16	18	60	84	56	250656
December 1943	0	0	14	15	34	63	88	67	250723
January 1944	0	11	24	24	34	93	93	93	250816
February 1944	8	21	28	26	12	95	95	95	250911
March 1944	5	24	21	13	23	86	95	86	250997
April 1944	4	23	11	25	41	104	95	104	251101
May 1944	0	4	25	24	47	100	95	100	251201
June 1944	0	7	20	25	23	75	75	75	251276
July 1944	0	11	16	17	20	64	58	64	251340
August 1944	0	1	3	2	0	6	9	6	251346

Chapter 3: Panzerkampfwagen Tiger Ausf.E

on 11 April 1942. Henschel mounted this first turret on their chassis **Fgst.Nr.V1**, completed in operational running order and examined on 15 April 1942.

As related by Dr. Aders in February 1945:

After working day and night, the first operational vehicle was loaded with cross-country tracks onto a railcar and sent to Hitler's headquarters, where it was unloaded with a 75 ton steam crane and set in operation. The demonstration on 20 April 1942 was successful. Problems occurred in the cooling system because the electromagnetic clutches couldn't transfer the necessary torque and slipped so that insufficient air passed through the radiators, which overheated. Also, the brakes didn't work properly. They seized because of overheating while braking, or they were adjusted with too little play.

The second **VK 45.01(H)** (**Fgst.Nr.250001** with **Motor HL 210 P45 Nr.46052** without a turret) was delivered to the **Kraftfahrversuchstelle Kummersdorf** (automotive proving ground near Kummersdorf) on 17 May 1942 after having been driven a total of 25 kilometers by Henschel. Henschel asked Krupp to deliver the second turret on 23 May 1942.

Just like the Porsche Tiger, the Henschel Tiger was plagued by automotive problems which caused significant delays in meeting production goals. On 18 June 1942, Henschel explained to Herr Kroemer, head of the **Tiger-Programm** under Speer:

After clarification of the necessary design and assembly modifications revealed by driving trials, the following schedule is promised by Henschel:
Two Tigers in June, 13 in July and 10 in August (on condition that problems with the ARGUS brakes are corrected by 25 June 1942), or 15 in July and 10 in August (on condition that ARGUS brakes are in working order by 10 July). This means that, except for one Tiger, the schedule shortfall from May and June can be made up in July.

By 5 July 1942, the schedule prognosis for the **Pz.Kpfw.Tiger (H)** was reported as follows: *The promise to deliver 15 Tigers in July 1942 can't be fulfilled because of problems that have appeared in the transmission, steering gear, and brakes. At least 10 Tigers will be delivered in August. Probably an increase to 15 Tigers is possible.*

On 13 July 1942, even with the additional automotive problems, Henschel optimistically reported to the head of the **Tiger-Programm**: *Driving trials with **Geraet C 10 Nr. 2 und 3** (i.e., **Fgst.Nr.250002** and **250003**) have revealed functional deficiencies in the final drives, transmission, and exhaust system. Under these conditions we can't give an exact estimate of the number of Tigers that can be completed in July. However, in any case the demanded total of 144 **Geraet C 10** will be produced by the end of April 1943.*

Footnote: **Geraet C 10** was Henschel's internal code for the **Tiger (H)**, and **Geraet C 11** was Henschel's internal code for the **Panther**.

Added to the problem of producing automotively reliable chassis, the **Hauptauschuss Panzerkampfwagen** (Main Committee Armored Fighting Vehicles) demanded that Henschel continuously increase their maximum monthly output. In January 1942 the peak monthly output goal had been set at 10 (to be met by August 1942). This was increased in July 1942 to 15 (to be met by December 1942), then in August 1942 to 30 (to be met by December 1942), in November 1942 to 50 (to be met by May 1943), in January 1943 to 75 (to be met by September 1943), and finally in March 1943 to 125 (to be met by April 1944).

Henschel managed to complete eight **Pz.Kpfw.VI(H)** in August 1942, four of which were sent to the **Heeresgruppe Nord** near Leningrad by the end of the month. These didn't get into action until 16 September 1942 (refer to Chapter 9), as three of these first four Tigers had problems with their transmission. The excuse given for not meeting the production goals in September, October, and November 1942 was that the demand for replacement parts and modification of the steering gear and transmission made normal assembly work impossible.

By December 1942, Henschel had finally solved most of the startup problems and by early 1943 were able to exceed their production goals. Henschel bragged that they had completed 230 Tigers by the end of April 1943, exceeding the total production goal of 207 by 23. Henschel reported on 3 May 1943 that too few Tiger turrets were available because the first turret had burnt out, two turrets didn't have the emergency hatch and therefore can't be delivered, three turrets needed the inner tube for sealing the turret ring exchanged, and two turrets had been sent to the troops as replacements. By the end of May 1943, Henschel had completed the assembly of 280 series production Tigers and Wegmann had delivered 302 Turrets.

Henschel continued to meet their monthly production goals until hit by a bombing raid in October 1943. Loss of production, contributed to bomb damage, was estimated as a total of 79 Tigers (30 in October, 24 in November, and 25 in December 1943 below the production goals). This was practically sufficient Tigers to have completely outfitted an additional two **Tiger-Abteilungen**. To make up for some of these losses, 18 excess **Panzerbefehlswagen** were converted to normal **Panzerkampfwagen** in November 1943.

For the first time, a large number of sabotage cases were discovered in Tigers that had been issued to **s.H.Pz.Abt.501**. As reported on 19 November 1943, 10 to 12 cm long steel shavings had been found in the idler wheel mounts of 10 Tigers; investigations were still in progress.

Tiger I production peaked at a total of 104 completed in April and 100 in May 1944. Then, as shown in Henschel's production report, the Tiger I was gradually phased out as the Tiger II was being phased in.

3.3.2 COMPONENT PRODUCTION

3.3.2.1 Armor

Fried.Krupp A.G. in Essen was awarded contract SS 210-5813/42(H) dated 22 July 1942 to fabricate the armor components for the three **VK 45.01 (H) Versuchsfahrzeuge**. Krupp, Essen was already overburdened with contracts for manufacturing armor components for the **Pz.Kpfw.IV** and the **Pz.Kpfw.VI P** and therefore D.H.H.V. (Dortmund Hoerder Huttenverein) was contracted to produce armor components for the first 424 **VK 45.01(H)** ordered for the production series. D.H.H.V. was originally assigned serial number sequence 250001 to 250621 for identification of their **VK 45.01(H)** armor components.

Following cancellation of Porsche Tiger production, Krupp was contracted to produce the armor components for 230 **VK 45.01(H)** in November 1942. The number was increased to 320 when

Stand des Programmes „Pz" im Fahrzeugbau 1944

Benennung		Jan	Febr	März	April	Mai	Juni	Juli	Aug	Sept	Okt	Nov	Dez
Fahrgestell Tiger I mit Aufbauten	soll	93	95	95	95	95	75	21,10 45 87	9 Rep				
		819	914	1009	1104	1199	1274	1340	1349				
	ist	93	95	86	104	100	75	64	6				
		819	914	1000	1104	1204	1279	1343	1349				
Fahrgestell Tiger II mit Aufbauten	soll	3	5	6	12	25	30	45	100	120	120	140	140
		6	11	17	29	54	84	129	229	349	469	609	749
	ist	3	5	6	6	15	32	45	97				
		6	11	17	23	38	70	115	212				

Henschel's report showing the Tiger I being gradually phased out as Tiger II production increased.

Hauptdienstleiter Saur (under Speer) decided that 90 turrets produced for the **VK 45.01(P)** would be converted to hydraulic traverse and used for the **VK 45.01(H)**. Krupp was originally assigned serial number sequence 250622 to 251200 on 23 December 1942, then 250421 to 251165 on 6 March 1943, and finally 251421 to 251954 for identification of their **VK 45.01(H)** armor components produced under contract number SS 210-5813/42.

Four of the former **VK 45.01(P)** turret armor bodies were sent by Krupp to Wegmann on 26 and 27 November, followed by 10 more on 22 and 23 December 1942. The last of the former **VK 45.01(P)** turret components with the original drum-shaped cupola had been sent to Wegmann by 11 June 1943. Krupp also started delivering the first of their order for 230 new turrets with periscope cupolas in June 1943.

Further contract extensions were awarded to Krupp and D.H.H.V. as additional contracts were awarded to Henschel and Wegmann for Tiger I assembly. Altogether, by the end of June 1944 Fried.Krupp A.G., Essen manufactured 537 armor hulls and turret armor bodies (including 90 converted **VK 45.01(P)** turrets and three **Versuchsserie**) and D.H.H.V. completed 758 armor hulls and turret armor bodies, for a grand total of 1295.

In order to complete the final contract for assembly of the last 54 Tiger I, on 21 June 1944 Krupp was ordered to refurbish and convert 17 hulls and 11 turrets and D.H.H.V. 23 hulls and 14 turrets, Henschel 15 lightly damaged hulls, and Wegmann 9 lightly damaged turrets. An additional 15 turrets from Krupp and 6 turrets from D.H.H.V. were to be supplied from their replacement parts contracts. Apparently the extra armor components (a total of 55 instead of the needed 54) were ordered because 11 out of the 51 hulls already returned to Krupp and D.H.H.V. for conversion to the latest standards had proven to be unrepairable.

3.3.2.2 Engines

Maybach, Friedrichshafen, the inventor of the high-performance **Maybach HL 210 P45 Motor** didn't maintain statistics on the production of the **HL 210** separate from their **HL 230**. Altogether in their **HL 210/230** series, Maybach produced 153 in 1942 and 4346 in 1943. After production of an estimated 1785 **HL 230** engines from January through April 1944, Maybach was completely knocked out of engine assembly by a bombing raid in late April 1944. Destruction was so devastating that Maybach didn't get back to producing the **HL 230** until October 1944.

Auto Union, Chemnitz at their Siegmar-Werk took over **HL 230** production in April 1944 and delivered 4366 **HL 230** engines for the Panther and Tiger from April 1944 and 1945. A total of 219 **HL 230 P45** engines were delivered from Auto Union to Henschel for installation in the Tiger I in 1944.

3.3.2.3 Transmissions

Adlerwerke, Frankfurt am Main was the main assembly firm for the Maybach-designed **Olvar 40 12 16 Getriebe** installed in the Tiger I. Statistics were not maintained separately for the production of the A and B variants of the **OG 40 12 16** transmissions but were reported together as 130 in 1942, 875 in 1943, 819 in 1944, and 26 ending in March 1945, sequentially numbered from 1 through 1850. A second factory, Waldwerke Passau managed by Zahnradfabrik Friedrichshafen A.G., started production of the **OG 40 12 16** transmissions in early 1944 and had completed serial number 20117 by September 1944.

3.3.2.4 Steering Gear

Henschel & Sohn, Kassel was the inventor and sole producer of the double-radius **Lenkgetriebe L 600 C** steering gear for the Tiger I. Production occurred as a single series, sequentially numbered starting with 1.

3.3.2.5 Main Gun

The **8.8 cm Kw.K.36 L/56** guns were assembled by two firms D.H.H.V. and Wolf Buchau. A total of 1514 **8.8 cm Kw.K.36 L/56** guns were completed, test fired, and accepted by **Heereswaffenamt** inspectors between January 1942 and July 1944".

3.2.6 Telescopic Sights

Ernst Leitz G.m.b.H, Wetzlar, the sole producer of the binocular **T.Z.F.9b**, reported that they had manufactured 1253 of these telescopic gun sights in the period from January 1942 to March 1944. The **T.Z.F.9b** was replaced by the monocular **T.Z.F.9c** telescopic gun sights in March 1944, and Leitz had already delivered 4 **T.Z.F.9c** before the end of the month.

GUN & GUN SIGHT PRODUCTION FOR THE VK 45.01 (H) & (P)

Month	8.8 cm Kw.K.36 L/56	T.Z.F.9b	T.Z.F.9c
Jan42	2	0	
Feb42	2	12	
Mar42	17	20	
Apr42	22	30	
May42	28	40	
Jun42	21	0	
Jul42	28	30	
Aug42	22	12	
Sep42	9	30	
Oct42	20	20	
Nov42	19	45	
Dec42	24	52	
Jan43	16	75	
Feb43	4	75	
Mar43	28	70	
Apr43	34	30	
May43	43	70	
Jun43	80	50	
Jul43	85	50	
Aug43	109	62	
Sep43	91	53	
Oct43	94	66	
Nov43	99	110	
Dec43	107	102	
Jan44	63	100	
Feb44	67	43	0
Mar44	95	6	100
Apr44	95	0	129
May44	91		142
Jun44	59		119
Jul44	40		103
Aug44	0		50

3.4 MODIFICATIONS DURING PRODUCTION

As with all production series German Panzers, modifications were frequently introduced during the production runs. These modifications were prompted by a desire for:

- improved automotive performance,
- increased firepower,
- added protection,
- simplified design for easier manufacturing, or were forced by shortages.

The modifications are divided into those incorporated in the chassis and turret. Within each subgrouping by component type, the modifications are listed in the chronological order in which the changes first appeared on completed Tiger I leaving the assembly plants. Changes in the camouflage colors affected all of the various Panzers and therefore have been consolidated in Appendix F.

Modifications for the Tiger I identified by the specific **Fgst.Nr.** or **Turm Nr.** were announced in the **H.T.V.Bl.** (army technical orders bulletin), described in the D 656 series of manuals, or found in meeting minutes or correspondence. Original documentation for many modifications did not survive. Modifications which were discovered by close inspection of the surviving Tiger I and photographs have been identified only by the month in which they appear to have been introduced or deleted.

In rare cases, several months elapsed between the first appearance of a modification and the time that it was present on all newly produced Tiger I. This was due in part to "first in, last out" tendencies when older stockpiled parts were covered, buried, or made inaccessible by storing shipments of newer parts. The newer parts, being easier to obtain, were used first until their removal allowed access to the older parts.

3.4.1 MODIFICATIONS TO THE CHASSIS

The primary causes of automotive problems were identified by Dr. Aders in his report dated 6 February 1945 on the development of the Tiger:

*Without conjecture or overstatement, it may be stated that the **Tiger E** chassis and its components not only fulfilled but exceeded expectations. The unavoidable teething problems encountered at first did not belong to the chassis design but occurred mainly in components which were ordered and procured by Wa Pruef 6.*

*The **Tiger E** had problems with the engine, transmission, brakes, and rubber tires. These problems created great concern and required a long time to eliminate.*

It would be easy to confirm that failures and breakdowns at the front – which frequently led to voluntarily blowing up brand-new vehicles – should be considered to be due almost exclusively to deficiencies in procured components. The rest is to be blamed on the fact that mass production had to start immediately, before test results were available; a point that can't be stressed often and strongly enough.

3.4.1.1 Internal Automotive Improvements

As related by Dr. Aders, automotive problems had already occurred during the demonstration of the first Tiger for Hitler's birthday on 20 April 1942. Engine coolant had overheated (because the electromagnetic clutches in the fan drives were slipping) and brakes had seized (possibly due to being set with too little play).

By early July, problems had also appeared in the transmission, steering gear, final drives, and exhaust system; and problems with the brakes still hadn't been cured. Further automotive problems experienced by units in the field are related in Chapter 9. The following corrections were made to cure these automotive problems, listed in the order in which they were introduced in the production series:

- Modified the steering gear, starting with **Fgst.Nr.250011** in September 1942
- Modified the track adjusting rachet, starting with **Fgst.Nr.250026** in November 1942
- Installed an improved 8-speed transmission, starting with **Fgst.Nr.250037** in November 1942
- Stopped engine coolant leakage with better hose clamps, starting with **Fgst.Nr.250101** in January 1943
- Lengthened the operating rod for the inertia-starter because the original rod was 41 mm too short, starting with **Fgst.Nr.250101** in January 1943
- Introduced the **Argus-Lenkapparat L.St.0.2** (steering device) with a smaller arc supported by two spokes in place of the Henschel **Osilit** steering wheel with four spokes by February 1943
- Changed track brake mounting as well as the entire brake lining, starting with **Fgst.Nr.250151** in March 1943. Originally the brakes were to be operated hydraulically. This arrangement presented significant problems and had to be abandoned in favor of mechanical operation. Since then, the brakes were brought up to a generally satisfactory working condition.
- Increased spring tension from 17.3 to 31.7 kg because the former tension on the clutch linkage was too weak, starting with **Fgst.Nr.250160** in March 1943
- Replaced oil lines to the fan drive system with asbestos-covered lines because oil was cooked out of the lines in the engine compartment, starting with **Fgst.Nr.250205** in April 1943
- Strengthened the main shaft in the steering gear, starting with **Fgst.Nr.250281** in June 1943.
- Installed a gasket between the exhaust pipe and muffler, starting with **Fgst.Nr.250385** in July 1943
- Installed a fuel filter, starting with **Fgst.Nr.250486** in September 1943
- Installed a wider rubber seal to prevent the loss of oil between the fan blades and body of the engine cooling system, starting with **Fgst.Nr.250505** in September 1943
- Incorporated all of the necessary improvements in the OG 40 12 16 transmission, starting with Nr.578
- Replaced the retaining pin in the drive wheel with two massive studs because the smaller pin had broken, starting with **Fgst.Nr.250675** in December 1943
- Replaced the felt ring with two labyrinth ring seals to prevent heavy oil loss from the final drives, starting with **Fgst.Nr.250689** in December 1943
- Changed the labyrinth seals in the roadwheel arms to improve lubrication retention, starting with **Fgst.Nr.250745** in January 1944
- Installed three grease fittings in the idler wheel arm housing so that it could be thoroughly lubricated, starting with **Fgst.Nr.250810** in January 1944
- Increased the filler pipe diameter to 33 mm to prevent oil from cooking out of the engine cooling fan drives, which could have caused fires, starting with **Fgst.Nr.250935** in March 1944
- Installed an overflow pipe on both sides of the engine cooling fan drives, starting with **Fgst.Nr.251045** in April 1944
- Installed an additional spring in the clutch foot pedal linkage since the already strengthened return spring was still inadequate, starting with **Fgst.Nr.251050** in April 1944
- Relocated the engine coolant temperature regulator from the firewall to the driver's position and added a fine adjustment, starting with **Fgst.Nr.251166** in May 1944
- Dropped the fourth heat sensor and spray head by the carburetor for the automatic fire extinguisher system, starting with **Fgst.Nr.251185** in May 1944
- Changed the central drive gear from beveled to straight teeth, starting with transmission Nr.1581 installed in **Fgst.Nr.251344** in August 1944

3.4.1.2 Periscopes in Hatches

Starting with **Fgst.Nr.250001** (delivered for testing by Wa Pruef 6 in May 1942) periscopes with protective armor hoods were mounted in the driver's and radio operator's hatches. These were not present on **Fgst.Nr.V1**, completed in April 1942.

3.4.1.3 Track Guards

After the requirement for a pivoting armor guard was dropped, the side extensions to the armor glacis over the tracks were cut off and the holes for the pivoting arms in the hull side extensions were cut away or plugged. Bent sheet-metal track guards were mounted on hinges at the front and frames were welded to the hull rear to support the rear track guards with three reinforcing ribs (starting with **Fgst.Nr.250001** in May 1942)

Mud, dust, dirt, and debris were thrown onto the top of the Tiger I by the 725 mm wide cross-country tracks extending beyond the pannier sides. Starting in November 1942, sheet metal mud guards were bolted to the pannier sides (and removed during rail transport). Initially, each of the four sections of the side track guards was made by welding an upper retaining strip to a lower guard strip.

After **Fgst.Nr.250028**, at or before **Fgst.Nr.250031** in November 1942, each section of the side track guards had been bent into shape using a single piece of sheet metal. At the same time the front and rear track guards were replaced by a new design with hinged side extensions (folded up for rail transport).

In December 1942 the four side track guard sections were mounted in a straight line (instead of following the bend in the lower edge of the superstructure side). By **Fgst.Nr.250085** in January 1943, triangular support pieces were added to the ends of the side track guard sections.

In May 1943, the frames, welded to the hull rear as a base for the rear track guards, were dropped. They were replaced by new hinged rear track guards (with cut-outs for the convoy light), supported by three hinge-plates welded directly to the hull rear.

3.4.1.4 External Equipment

No mounts or fasteners for tools or equipment were welded to the **Fgst.Nr.V1** and **250001** delivered to Wa Pruef 6 for testing in April and May 1942. Then the pattern for stowing tools and equipment on the outside to the Tiger I hull changed continuously throughout the production run. Some of the changes, noted in

Chapter 3: Panzerkampfwagen Tiger Ausf.E

orders or observed in photographs or on surviving specimens, include:

- Discontinued the horn mounted on the deck to the right of the driver's hatch (starting with **Fgst.Nr.250001** in May 1942).
- Welding fasteners on the deck for five cleaning rods (four 1238 mm long, one 1240 mm long – two on left side, three on right side) and two wrecking bars (starting in August 1942 in accordance with drawing J2802 dated 22 July 1942)
- Welding two tubes to store the engine hand-cranked starter guide plate on the hull rear between the mufflers (by August 1942, repositioned for vertical stowage of the starter guide plate in December 1942)
- Mounting a storage tube for a spare antenna rod along the outer edge of the superstructure side at the right rear and a holder for a fire extinguisher on the rear deck (starting in August 1942)
- Welding fasteners to the left superstructure side to store the **15 meter** long cable used for replacing tracks (starting by **Fgst.Nr.250011** in September 1942)
- Welding fasteners for securing a hammer, spade, axe, wooden jack block, wire cutters, wrecking bar, two towing cables, and a fire extinguisher on the deck, a 15-ton jack on the hull rear, and tabs for securing a tarp over the engine deck (by **Fgst.Nr.250011** starting in September 1942)
- Replacing the rectangular combination stop/convoy light mounted on the rear with a cylindrical convoy light in October 1942
- Reversing the location of the tow cable fasteners on the deck and mounting the 15-ton jack horizontally on the right hull rear in October 1942
- Welding fasteners on the glacis to secure a long handled shovel by **Fgst.Nr.250031** in October 1942, discontinued by January 1944)
- Extending the hull side at the front, allowing the U-shaped tow shackles to pivot freely (by **Fgst.Nr.250055** in December 1942)
- Mounting a track tool stowage box on the hull rear above the left rear track guard in December 1942
- Moving the tow cable eye fasteners and antenna stowage rod when the **S-Minen** discharger mounts were added in late December 1942
- Welding fasteners on the lower left hull rear for securing a shaft for the engine hand-cranked starter (by **Fgst.Nr.250122** in February 1943)
- Changing the left brackets to hold three (instead of two cleaning rods) and relocating these brackets for securing shorter gun cleaning rods (one 918 mm long, five 980 mm long) (by **Fgst.Nr.250122** in February 1943)
- Welding redesigned fasteners for stowing the track replacement cable on the superstructure side (by **Fgst.Nr.250461** starting in August 1943)
- Welding a tab onto the glacis to secure the chain holding a plug for the machine-gun ball mount starting in August 1943
- Mounting C-hooks on the left hull rear and on the deck to the right of the driver's hatch (by **Fgst.Nr.250496** starting in September 1943)
- Welding fasteners on the deck by the antenna for stowing a starter crank handle (by **Fgst.Nr.250570** starting in October 1943)
- Discontinued mounting the track tool box on the left hull rear in October 1943.
- Issuing a new U-shaped tow shackle (starting with **Fgst.Nr.250675** in December 1943)
- Modifying the mounting on the right hull rear for supporting a sturdier 20-ton jack (starting with **Fgst.Nr.250772** in January 1944)
- Dropping the fasteners holding the shovel across the glacis starting in January 1944
- Cutting the face of the hull side extensions at the front, drilled for mounting tow shackles, to allow additional room for fitting C-hooks (starting in January 1944). The hull side extensions at the rear, drilled for mounting two shackles, were also reshaped with a lower lip.
- Relocating tool stowage on the deck after a turret ring guard was added (starting with **Fgst.Nr.250850** in February 1944)
- Adding mounting bolts to the engine hand-cranked starter guide plate starting in February 1944
- Issuing a new U-shaped tow shackle in which the pin was retained by a C-shaped key held by a cotter key (starting with **Fgst.Nr.250875** in February 1944)

3.4.1.5 Provision for Submerged Fording

Starting in August 1942, a hinged flapper was installed on top of each exhaust muffler to prevent water from flowing in during submerged fording. As a seat for a sealing cap, a groove had been cut into the armor cover for the machinegun ball mount. By September 1942, two hinged toggle bolts had been welded to the driver's front plate, one on each side of the ball mount armor cover, to secure the sealing cap. These toggle bolts were dropped by June 1943, but the groove was still cut into the ball mount armor guard until the end of the Tiger I production run.

Many additional gaskets and seals were needed to make the Tiger I watertight, as revealed in the order dated 30 August 1943 from Wa Pruef 6 to drop the submersibility requirement in an effort to make the Tiger easier to produce:

*Submersibility of the **Pz.Kpfw. Tiger I** is to be dropped immediately. However, fordability up to at least 1.50 meters is required. Installation of seals and preparation for these seals, needed for diving, can stop immediately.*

The following parts can be dropped from the turret: cap for the ventilation fan with rubber gasket and bolts, seal for the gun mantlet, plugs for the machinegun opening, seals for the openings in the gun mantlet for the gun sight, inflatable rubber tube in the turret race and the special muzzle cap for the gun. Seals for the commander's hatch, the glass vision blocks, the loader's hatch, the pistol ports, and the escape hatch are to be retained to prevent rain water entry.

The following parts can be dropped from the chassis: seal for the air intake cap on the deck above the transmission, complete seal for the firewall, gasket under the engine hatch including tightening the hatch cover, air intake cap on the engine hatch, cover for the air intake slit in the engine hatch, hinged flaps on the mufflers, four-part telescoping snorkel with accessories, snorkel ductwork in the engine and crew compartment, ventilation louvers and the operating cables in the left and right panniers, butterfly valves for ventilation of the engine exhaust headers, and seals in the partition between the engine compartment and coolant system in the side panniers.

The seals for the driver and radio operator hatches and periscopes are to be retained to prevent rain water entry. To meet the requirement for fording up to a depth of 1.50 meters, all other seals and gaskets (including those for the driver's visor, machinegun ball mount, and the entire suspension) are to be retained.

This order to drop the parts needed for submerged fording went into effect starting with the 496th Tiger chassis completed by Henschel, **Fgst.Nr.250493** in September 1943. Only the first 495 Tigers were partially outfitted for submerged fording. The hinged flapper on top of each exhaust muffler was no longer mounted by October 1944.

Changes to the ductwork on the hull floor, deleting the rear duct in the engine compartment and the diversion plug, were not implemented until **Fgst.Nr.250625** in November 1943. The bilge pump (underneath the turret drive with its associated drive, controls, and discharge pipe) was not dropped until **Fgst.Nr.250762** in January 1944.

3.4.1.6 Drive Sprocket

The spokes in the drive sprocket wheel were originally aligned with the sprocket ring teeth and bolts. Starting in August 1942, the drive sprocket ring was realigned so that the spokes were centered between teeth.

The hub of the drive sprocket was machined flat in order to better secure the five bolts with reinforced tabs (starting with **Fgst.Nr.250220** in April 1943).

Additional double-locking stripes, shared by two nuts, were introduced for securing the toothed sprocket ring, because differences in castings did not always result in centered locations for securing the nuts (starting with **Fgst.Nr.251205** in June 1944).

3.4.1.7 Rear Deck

Concerned that the armor louvers over the cooling fans would be too restrictive to air flow, initially these could be opened remotely with hydraulic cylinders. This system was quickly abandoned. By mid-August 1942, pivoting hooks and catches were welded on to hold the rear armor louvers open while working on the fans. A pivoting hook and catch were also added to secure the engine hatch when opened.

Due to the requirement that the Tiger be able to ford across bodies of water up to 4 meters deep, the engine compartment was completely sealed off except to let in engine combustion air. During normal driving, combustion air entered through a circular opening (protected by an armor cap) on the engine hatch. Then, because fumes from fuel leaks built up causing fires in the engine compartment, a rectangular slit was cut across the front end of the engine hatch by August 1942. A rectangular plate (to seal the air intake slit in the engine hatch during submerged fording) was secured on the deck to the left of the turret (by **Fgst.Nr.250122** in February 1943).

Starting in March 1943, the cap over the ventilation riser was dropped and a "triangular" shaped opening (with cover plate held by three bolts) added to allow access for engine component adjustments without needing to open the large engine hatch. To simplify casting, the design of the right rear armor grating for cooling air exhaust was changed to match the design of the left rear armor grating, starting in April 1943. A pivoting **Anschlag** (stop) was mounted on both sides to support a partially opened engine hatch (starting in late April 1943) directly followed by relocating the handles. A forged armor cap (instead of a welded cap) covering the combustion air intake on the engine hatch was introduced by Krupp by August 1943. The seal inside the armor cap covering the combustion air intake on the engine hatch was discontinued in May 1944.

3.4.1.8 Camouflaged Like a Truck

To mount a tubular frame as the forward support for a canvas cover to camouflage the Tiger to appear like a large truck, holes were cut into glacis plate at the far left and right corners below the driver's front plate (in August 1942). Reinforcing rings were added to create a level base (by **Fgst.Nr.250053** in December 1942).

3.4.1.9 Wintermassnahmen (Winterization)

Devices to aid in starting in cold weather, including a **Anlasskraftstoff-Einspritzvorrichtung** (starting fluid spray injector) and **Kuehlwasseruebertragung** (heated coolant transfer), were to be installed in Tiger I already completed in August 1942.

On 10 September 1942, Henschel was directed to furnish a **Kampfraumbeheizung** (crew compartment heater) for the Tiger. The **Kampfraumbeheizung** consisted of a sheet metal housing fitted over the cooling air exhaust grating on the left side. Air warmed by the left radiator was diverted by the **Kampfraumbeheizung** into the snorkel pipe opening on the rear deck and on through ductwork which fed the warmed air into the crew compartment.

All Tigers sent to **Heeresgruppe Sued** from December 1942 to February 1943 had been outfitted with a **Kampfraumbeheizung**. But, as reported by the Maybach representative at the front, all of the **Kampfraumbeheizung** had been removed from the Tigers to avoid causing additional engine fires.

The **Fuchsgeraet** (coolant heater using a blow torch) was first installed in the Tiger I starting with **Fgst.Nr.250823** in February 1944. The **Fuchsgeraet**, connected into the engine coolant flow path, was mounted on the left side of the engine. The blow torch was suspended on a hanger on the hull rear. Flames from the blow torch were directed into the **Fuchsgeraet** through a hole cut into the hull rear below the left muffler armor guard. A flattened chimney directed heat from the blow torch flames upward to prevent ignition of fuel fumes inside the engine compartment.

As announced on 1 June 1944 in the **H.T.V.Bl.**, a **Kampfraumbeheizung** (crew compartment heater) wouldn't be installed in **Pz.Kpfw.Tiger I**.

3.4.1.10 Radio Equipment

An extension for mounting an antenna (located on the upper right corner of the hull rear) was cut off and the rectangular slit for the antenna wire in the hull rear filled with a welded plug (by **Fgst.Nr.250014** in October 1942).

To reduce damage to the radios from hits, the radio racks were strengthened (starting with **Fgst.Nr.250875** in February 1944). The **Entstoerer EM/S 100/L** (interference suppressor) was replaced with **EM/S 75/1**, except that the **EM/S 100/L** was retained in **Pz.Bef.Wg.** (starting with **Fgst.Nr.250985** in March 1944).

Because vacuum tubes in the **Kasten Pz.20** (intercom box) broke when shaken too violently, both the box and incoming cable were spring-mounted (starting with **Fgst.Nr.251051** in April 1944).

3.4.1.11 Feifel Air Filters

As prefilters for engine combustion air, pairs of **Feifel** air cleaners were mounted on the left and right upper corners of the hull

Chapter 3: Panzerkampfwagen Tiger Ausf.E

rear (starting in November 1942, mounting plates had been installed by **Fgst.Nr.280014** in October 1942). Air was drawn in through flexible ducts on the engine deck, passed through the prefilters, and then through flexible ducts to a slit cut into the engine hatch, where it entered the engine compartment. Except in dust-free conditions, the round armor cap over the engine combustion air intake vent was to be kept closed as long as the prefilters were in operation. **Feifel** prefilters with a simplified upper chamber design were introduced in March 1943. **Feifel** prefilters with the brackets for hoses and ductwork on the engine deck were discontinued in October 1943. The air intake slit in the forward end of the engine deck was then protected by an elevated plate. However, the small mounting plates were still welded onto the top of the hull rear into December 1943.

3.4.1.12 Gleisketten (Tracks)

Pz.Kpfw.Tiger with **Fgst.Nr.250001** to **250020** were outfitted with right and left cross-country tracks. Starting with **Fgst.Nr.250021** in October 1942, standardized **Kgs 63/725/130** cross-country tracks were fitted to both sides.

Because the track pins had drifted out on their own, new track pins retained by a ring and **Hettmannstift** were introduced, starting with **Fgst.Nr.250145** in March 1943.

As announced in the **H.T.V.Bl.** in March 1944: *All new production armored full-tracked vehicles (other than the Pz.Kpfw.38 and its variants) receive tracks with **Gleitschutzpickeln** (skid preventing chevrons) cast onto the track face. These tracks are to be mounted on the Panzer only directly before they are loaded on rail cars in order to preserve roads.* Cross-country tracks with **Gleitschutzpickeln** had already been issued to Tiger I completed by Henschel as early as **Fgst.Nr.250570** in October 1943.

Following the introduction of the smaller diameter idler (600 instead of 700 mm), the track pin return plate (mounted on both hull sides above the idlers) was improved by being widened and mounted at an angle starting in May 1944.

3.4.1.13 Penetrations in the Belly Plate

A drain valve in the belly was deleted (starting with **Fgst.Nr.250051** in December 1942).

When the more powerful **Maybach HL 230 P45** engine was introduced, a large rectangular hole (200 x 535 mm) for access to the electrical generator and fuel pumps was cut into the belly plate in place of the left-hand 250 mm diameter hole (starting with **Fgst.Nr.250251** in May 1943). Another fuel drain port was added on the right side.

A cover box was added to protect the handwheel for the drain valve on the floor by the radio operator (starting with **Fgst.Nr.251075** in April 1944).

The cable-controlled drain valve in the hull was changed because the older design with a spring had allowed water to seep in (starting with **Fgst.Nr.251165** in May 1944).

3.4.1.14 Exhaust Muffler Guard and Deflector

From experience it was learned that the glowing exhaust muffler and flaming exhausts were visible, especially in snow-covered terrain. The open sides of the mufflers were surrounded by sheet metal guards, and at a suitable clearance the exhaust was deflected by a round cover plate supported by five rods (started by **Fgst.Nr.280082** in January 1943).

3.4.1.15 Reinforced Roadwheels

Tabs holding roadwheel bolts had torn off or vibrated loose. Locking strips were introduced spanning two bolts (starting with **Fgst.Nr.250101** in January 1943).

The original six large bolts proved to be ineffective in retaining the steel rim (next to the track guide teeth) on each roadwheel. An interim solution to the problem had been introduced by February 1943 when two smaller diameter bolts were added as reinforcement for each original bolt (for a total of 18 bolts retaining the steel rim on each roadwheel). The final solution of securing the steel rim with 12 large bolts was introduced starting with **Fgst.Nr.250220** in April 1943.

The conical roadwheel disc had failed at the inner radius weld seams. The base of the new discs were fastened with bolts and reinforced by a weld bead (starting with **Fgst.Nr.250314** in June 1943).

3.4.1.16 Driver's Vision Devices

The **K.F.F.2** (twin driver's periscopes) were no longer installed (by **Fgst.Nr.250122** starting in February 1943). Until hulls became available which hadn't been drilled for the twin periscopes, the twin holes in the driver's front plate were plugged and welded shut.

The five-spoke knurled handwheel for opening and closing the driver's visor was replaced with a three-spoke smooth handwheel by August 1943.

3.4.1.17 Zerstoererpatronen (Destruction Charges)

Entry 146, in the AHM (general army bulletin) on 8 February 1943, advised the troops that all armored vehicles were to be outfitted with **Sprengpatronen Z** for destruction of equipment and ammunition. Following a meeting on 13 April 1943, Wa Pruef 6 reviewed a copy of Henschel's drawing of holders for **Zerstoererpatronen**. Guidance on the use of the **Zerstoererpatrone Z 85** for blowing-up equipment to prevent its capture was published in July 1943 as follows:

*It is forbidden to allow a repairable **Pz.Kpfw.VI "Tiger"** to fall into enemy hands. The **Zerstoererpatronen** must always be carried in the Tiger, secured in holders on the cross-member above and behind the driver's and radio operator's seats. Under all circumstances, it must be ensured that the same components are destroyed in all Tigers so that it will be impossible for the enemy to repair a Tiger by retrieving parts from several captured Tigers. To achieve this one **Zerstoererpatrone Z 85** is to be placed inside the breech of the main gun to destroy the gun and the interior of the fighting compartment. A second **Zerstoererpatronen Z 85** is to be placed on the engine to destroy the engine and engine compartment.*

3.4.1.18 Maybach HL 230 P45 Engine

The **Maybach HL 230 P45** engine was installed, replacing the **Maybach HL 210 P45** engine (starting with **Fgst.Nr.250251** in May 1943). New fan drives with twin drive shafts and friction clutches had to be adopted with the new engine. Since the hole in the hull rear was no longer centered for the new engine mounting, another hole was drilled in the **Schwungkraftanlasser** (inertia hand-cranked starter) guide plate to line up the shaft for the new motor.

A new piston design was installed in the **Maybach HL 230 P45** motors (starting with serial number 932338), reducing the compression ratio from 1:68 to 1:64. Problems with blown head gaskets were corrected by installing copper rings pressed into grooves to seal the heads of **Maybach HL 230 P45** motors, starting with serial number 932453 in August 1943. Other modifications, introduced at the same time, included improving coolant circulation inside the engine and installing a reinforced membrane spring in the fuel pump.

In November 1943, starting with **HL 230 P45** motor number 932723, the governor was preset at the factory for a maximum speed of 2500 rpm under full load and the motors were outfitted with a hand-operated temperature control on the oil cooler.

After reduction of the engine speed to 2500 rpm, the turning radius for each gear and the maximum speed was reduced to:

Gear	Smaller Curve	Larger Clutch	Speed at 2500 rpm
1	3.5 m	11 m	2.5 km/hr
2	5.5 m	17	3.5
3	8	24	5.0
4	12	35	7.5
5	18	54	12
6	27	80	18
7	39	116	25
8	57	173	38
R	4.7	14	

Maybach HL 230 P45 motors with serial numbers from 932593 to 932828 and motors that were rebuilt in October and November 1943 (with M stamped on the serial number plate) had faulty bearings that frequently failed. Improved bearings were installed in **HL 230 P45** motors starting in January 1944.

On 7 March 1944, units were advised that some of the **Maybach HL 230 P45** motors had been outfitted with a **Durchdrehanlasser** (reduction gear hand-cranked starter) instead of a **Schwungkraftanlasser** (inertia hand-cranked starter).

3.4.1.19 Shock Absorber Mounting

The mounting bolt for the shock absorbers was changed to one with a large external head because the previous conical shaft design had vibrated loose (starting with **Fgst.Nr.250301** in June 1943).

3.4.1.20 Fuel Tanks

The mountings and fasteners for the fuel tanks were improved (starting with **Fgst.Nr.250351** in July 1943). The diameter of the filler to the upper fuel tanks was increased from 72 to 117 mm (starting with **Fgst.Nr.251165** in May 1944). The upper fuel tanks on the left and right were modified (starting with **Fgst.Nr.251201** in May 1944).

3.4.1.21 Scheinwerfer (Headlight)

The number of **Scheinwerfer** (headlights with blackout driving slits) was reduced to a single headlight mounted on the deck to the left front of the driver's hatch (starting in August 1943). The single **Scheinwerfer** was relocated to a position in the center of the driver's front plate (by **Fgst.Nr.250570** in October 1943).

On 19 October 1943, OKH ordered: *The new type of removable **Scheinwerfer**, mounted on all types of armored vehicles, to be removed from the holders and stored securely prior to entering combat and for the duration of the action.*

3.4.1.22 Motortrennwand (Firewall)

The firewall separating the engine compartment from the crew compartment was replaced by a new two-piece design (starting with **Fgst.Nr.250501** in September 1943). A new fan housing was installed in the firewall (starting with **Fgst.Nr.250625** in November 1943).

Five electrical components (**Reglerschalter RS/KN600/70 12/1, Entstoerer EM/S 100/1, Schaltschuetz SH/SE 8/1, Sammlerhauptschalter SSH 68/1 Z**, and **kleiner Entstoerer EM S 5/1**) were moved out of the engine compartment and mounted on the crew compartment side of the firewall (starting with **Fgst.Nr.250861** in February 1944)

3.4.1.23 Electrical Equipment

An electrical heater surrounding the batteries was installed (starting with **Fgst.Nr.250696** in December 1943). Removing the starter key did not disconnect power to the battery heater after switch box **HAW 12/1** was installed (starting with **Fgst.Nr.25095** in March 1944). This problem was corrected by installing standard switch box **HAW 12/2** with two keys, replacing **HAW 12/1** (starting with **Fgst.Nr.251165** in May 1944).

A rubber protective sleeve was installed for the battery cable (starting with **Fgst.Nr.251150** in May 1944). Cables penetrating the firewall were protected to prevent electric shorts (starting with **Fgst.Nr.251336** in July 1944).

3.4.1.24 Gummigefederten Stahllaufrollen (Rubber-Cushioned Steel-Tired Roadwheels)

Starting with **Fgst.Nr.250822** in February 1944, **Pz.Kpfw.Tiger, Ausfuehrung E** were outfitted with **gummigefederten Stahllaufrollen** (rubber-cushioned, steel-tired roadwheels). The following changes were introduced in comparison to the previous suspension:

• Tracks, roadwheel arms, ball-bearing race, and labyrinth seals were the same for both models.
• The hub caps were not interchangeable between models.
• There were only two instead of the previous three roadwheels per axle. With only two roadwheels per axle, the outer roadwheel was not to be removed for loading on rail cars.
• Steel tires were not interchangeable with rubber tires.
• Only in an emergency were roadwheels of different models to be used as replacements.

The troops were authorized to replace all of the normal rubber-tired roadwheels on **Pz.Kpfw.Tiger** with the new **gummisparendem Laufrollen** (rubber-saving roadwheels).

To prevent the bolts holding the hub caps on the **gummigefederten Laufrollen** (rubber-cushioned roadwheels) from working loose, the bolt size was increased from M10 to M12 (starting with **Fgst.Nr.251025** in April 1944).

Chapter 3: Panzerkampfwagen Tiger Ausf.E

4.1.25 Smaller Diameter Idler Wheel

After changing over to the steel-tired roadwheels, a smaller diameter (600 mm instead of 700 mm) was introduced starting in February 1944 (possibly for better self-cleaning as was the case with the Panther Ausf.G).

4.1.26 Wooden Decking over Fuel Tanks

Wooden decking was installed over the top of the upper fuel tanks to catch shell fragments and bullet splash coming down through the cooling grating (starting with **Fgst.Nr.251075** in April 1944).

4.1.27 Reparierte Wannen (Recycled Hulls)

The last 54 Tiger I, **Fgst.Nr.251293 – 251346**, were to be produced utilizing hulls recycled from Tigers that had been returned from the front for major rebuild. In a meeting on 14 June 1944, D.H.H.V., Krupp, and Henschel determined the changes that were needed to "modernize" these recycled hulls to meet the latest standards:

*For modernization, the following modifications to the armor hulls are to be accomplished by the **Panzerfirmen** (D.H.H.V. and Krupp):*

Kettenbolzenabweiser 021C2701-401 (track pin return plate) is now wider. Cut off the old plate and replace it with a new plate set at an angle.

Rueckwand 021B2701-10 (hull rear) Water drain holes, several recessed holes for the cooling system, and a long hole for an antenna have been deleted. Fill the unneeded holes. Bore a hole and weld on a flange for the **Fuchsgeraet** (coolant heater).

Boden 021B2701-365 (belly plate) The 250 mm diameter hole has been changed to a rectangular (200 x 535 mm) hole. A new hole, 120 mm in diameter, has been added. A drain valve opening has been deleted. Bore the new holes and weld the covers for the unneeded holes shut. Remove the rear ductwork and air intake tube from the engine compartment.

Trennwand 021C2701-370 (firewall) All cutouts, holes, and labyrinth seals have been moved. Cut out the old firewall and weld in the new two-piece firewall.

Stirnwand 021A2701-5 (front plate) The hull side extensions have been changed and a mounting for a headlight added. Shorten and cut out the hull side extensions and weld on the mounting for the headlight.

Klappe ueber Motor 021C2701-13 (engine deck hatch) The cylindrical holes have been deleted and the handles moved. Weld cut-off bolts in the cylindrical holes. Cut off the handles and weld them in the new location.

3.4.2 MODIFICATIONS TO TURRET AND ARMAMENT

Krupp shipped the first operational **VK 45.01(H)** turret **Nr.1** with **8.8 cm Kw.K.36 L/56 Rohr Nr.1** to Henschel by truck on 11 April 1942. Mounted on the first **VK 45.01(H) Fahrgestell Nr.V1** completed by Henschel, the turret was inspected and tested on 15 April 1942. Following this inspection, a lengthy list of modifications to the **VK 45.01** turret was discussed by representatives from In 6, Wa Pruef 6, Henschel, and Krupp on 15 and 16 April 1942:

*The new spring-cushioned machinegun mount with improved seating is to be installed in the **VK 45.01** turret. The same space as in the **Pz.Kpfw.IV** is to be allotted for the belt feed and attaching both belt bags.*

To prevent the hatch lid for the commander's cupola from slamming shut during cross-country travel, it is to be modified so that a bump stop and retainer hold the lid open at about 110°.

The commander's view out of the turrets now in production will be improved by widening the inside of the vision slits in the outer armor ring. This will allow a field of view from the individual vision slits starting from about 15 meters out. The armor plates now installed inside the vision blocks are to be deleted and the head cushions will be bolted directly to the frames for the vision blocks. The frames are to be reinforced and introduced into production as soon as possible.

A counterweight is to be installed for the loader's hatch lid in the turret roof. The lid is to open as far as possible, at least 110°, and the counterweight located so that it does not interfere further with the loader performing his tasks.

The footrests for the commander are to be changed so that the pegs now used are replaced by a plate that is slanted somewhat to the rear.

*An additional crew hatch can't be installed in the side or rear of the first 100 turrets for either the **VK 45.01 (H)** and **(P)** series because of production reasons.*

Vision slits are present in the turret sides for both the gunner and loader. In addition, periscopes facing forward are to be mounted in the turret roof for both the gunner and loader.

The height of the loader's space is uncomfortably low at about 1.60 meters. Because of the batteries, the turret platform can only be lowered a maximum of about 25 mm. Because this measure will not result in significant improvement in comparison to the difficulties caused to series production, designs to modify the turret platform will be abandoned. The reinforcing ring around the turret platform is to be set so deep that the upper rim is even with the platform.

Krupp is tasked with mounting an auxiliary turret traverse mechanism to be used by the commander or loader.

Krupp will attempt to move the azimuth indicator forward so that it is comes into the gunner's field of vision through a slight movement of his head. The gunner's seat is to be lowered by 40 mm so that it is easier to turn the elevation handwheel.

*Relocating the traverse from the left hand to the right and elevation from right hand to the left is not possible because of the overall design of the **VK 45.01** turret traverse and gun elevation mechanisms. A short handgrip is planned to be added to the handwheel for the elevation mechanism. Even though desirable, it isn't possible to increase the diameter of the handwheel because it will hit the cradle when the gun is elevated.*

The lever for electrical firing is to be bent so far to the right that the space between the lever and the handwheel is at least 60 mm.

*Henschel is to investigate mounting an external travel lock similar to the type used on the **8.8 cm Flak** that supports the gun as close to the muzzle as possible. The gun can be traversed to the rear during road marches.*

The opening for the turret ventilator is to be shaped so that rainwater can't get into the turret.

About eight to ten machinegun cartridge bags are to be stowed in a niche next to the loader. The rest of the cartridge bags are to be stowed in the crew compartment so that they can be easily reached by the loader.

All five sets of cooking utensils are to be stowed in a niche in the superstructure.

On 1 May 1942, Krupp sent Wegmann the following list of modifications with drawing changes, that were to be made starting with **Turm Nr. 1**:

1. Change the **Ausgleicher** (counterbalance) arrangement
2. Change the **Kommandantenkuppel** (commander's cupola):
   a. Secure the opened **Deckel** (hatch lid) at 110°
   b. Widen the **Sehschlitze** (vision slits)
   c. Delete the **Panzerplatten** (armor plates) and their mountings from the glass vision blocks
   d. Modify the **Rahmen** (frames) for the vision blocks
   e. Bolt the **Stirnpolster** (head cushion) to the frames
3. Lower the **Richtschuetzensitz** (gunner's seat) by 40 mm
4. Shorten the **Abfeuerbuegel** (firing lever)
5. Lengthen the slit in the arm for the **Geschuetzzurrung** (gun travel lock)
6. Secure the bolts holding the inner **Dichtring** (sealing ring)
7. Modify the **Ladesitz** (loader's seat) so that it can be folded up
8. Widen the ribs in the console for the **Richtmaschine** (traversing mechanism)
9. Modify the lifting arm for the **Lukendeckel** (loader's hatch lid)
10. Change the hand grip on the **Seitenrichthandrad** (traversing hand wheel)
11. Modify the bridge for mounting the **Optik** (gun sight)
12. Add a stop for the **Turmzurrung** (traverse travel lock)
13. Modify the backrest for the **Richtschuetzensitz** (gunner's seat) so that it can be folded up
14. Replace the footrest pegs for the commander with a **Trittblech** (step plate) sloped 30° downward
15. Lower the **Drehbuehenrand** (edge for the turret platform)
16. Add a rainwater drain for the **Luftaustriffsoeffnung** (ventilation opening)
17. Lower the **Kommandantensitz** (commander's seat) by 30 mm

Additional changes, such as adding an auxiliary turret traverse drive and mounting periscopes in the turret roof, are to occur in the future following completion of the supporting drawings.

3.4.2.1 Nebelwurfgeraet (Smoke Candle Dischargers)

In June 1942, Wa Pruef 6 provided instructions for mounting triple **Nebelwurfgeraet** (smoke candle dischargers) facing forward on each turret side. Although **Nebelwurfgeraet** were not yet available, starting in August 1942 Wegmann prepared the turrets by welding the mounting brackets to the turret sides and drilling holes for firing circuit wiring so that the troops could mount **Nebelwurfgeraet** themselves. Complete **Nebelwurfgeraet** were mounted at the assembly plant starting in October 1942.

Troops at the front reported that **Nebelkerzen** (smoke candles) hit by small-arms fire were set off in the **Nebelwurfgeraet**. In calm wind conditions, the smoke enveloped the Tiger, causing a loss of visibility and sometimes incapacitating the crew. Because of this hazard, **Nebelwurfgeraet** were no longer mounted on the turret sides (starting with **Turm Nr.286** in June 1943).

3.4.2.2 Turmzurrung (Turret Traverse Lock)

On 19 August 1942, Wa Pruef 6 reported: *For the present the expedient method invented by Wegmann for securing the turret traverse lock with a locking pin will be retained. As a further modification, Krupp has strengthened the spring retaining the travel lock pin. A toothed segment travel lock is planned for later model.*

A report from the front stated: *The pin for the turret travel lock repeatedly falls out due to vibration of the vehicle. When the vehicle is not level, the turret swings around by itself, resulting the gun tube striking and knocking shut the driver's and radio operator's opened hatch lids.*

The cylindrical pin for the turret traverse lock was replaced by a conical pin (starting with **Turm Nr.113** in February 1943). toothed-segment travel lock was introduced with the new turret design (starting with **Turm Nr.392** in July 1943).

3.4.2.3 Gepaechkasten (Baggage Bin)

Starting in August 1942, a temporary solution to the need for stowing crew baggage had been met by mounting a modified **Pz.Kpfw.III** stowage bin on the turret rear of the **Pz.Kpfw.VI(H)** (With the exception of some Tigers issued to the **1.Kp s.H.Pz.Abt.502** and **s.H.Pz.Abt.503**, the smaller **Pz.Kpfw.III** style stowage bins were replaced by larger bins prior to the units being sent to the Front.)

A contract had been awarded to Vorrichtungs und Geraetebau GmbH to design a **Gepaechkasten** specifically for the **VK 45.0** turrets by 25 June 1942. On 21 November 1942, a decision was made to speed up production of the new **Gepaechkasten** constructed in accordance with drawing AKF31552. A sample **Gepaechkasten** for **Pz.Kpfw.VI Ausf.P1 u. H1** was sent by Vorrichtungs und Geraetebau to Krupp, Essen on 5 January 1943. After the sample was examined by Wa Pruef 6, a decision was made on 21 January 1943 to mount this **Gepaechkasten** on a **Pz.Kpfw.VI(H) Ausf.H1** turrets. Mounting the **Gepaechkasten** close to the turret rear displaced 6 of the 15 hangers for spare track links that had been planned to be mounted between the escape hatch and the pistol port around the turret rear.

Vorrichtungs und Geraetebau GmbH delivered the first batch of 15 **Gepaechkasten** by the end of January and the second batch of 15 on 5 February, in time to start outfitting **Pz.Kpfw.VI(H)** with **Gepaechkasten** at the assembly plant starting in late January/early February 1943.

3.4.2.4 Federausgleicher (Spring Counterbalance)

A spring counterbalance for the loader's and commander's hatch lids was added (starting with **Turm Nr.30** in November 1942)

3.4.2.5 Regenschutzhaube (Rain Shield)

A rain shield for the commander was added (starting with **Turm Nr.30** in November 1942). This rain shield was made of a rectangular piece of canvas stretched over a wire frame and supported by two rods inserted into sleeves (initially pegs were used) welded onto the top of the cupola. The support sleeves welded to the cupola were mistaken by the Allies examining a captured Tiger as being supports for a range finder.

3.4.2.6 Notabfeuerung (Emergency Firing)

On 19 October 1942, Wa Pruef 6 asked Krupp to complete the design for installing an inductive emergency firing device so that it could be installed for the **8.8 cm Kw.K.** by the assembly arms starting in November 1942. It was possible for the gunner to depress the emergency firing switch with his knee without removing his hands from the elevating or traversing handwheels.

On 24 February 1943, Tiger crews in Africa reported: *The normal electrical firing circuit fails when the turret is hit by enemy fire. Then the inductive **Notabfeuerung** (emergency firing device) has to be used to fire the gun. Better protection against sudden shocks is needed.*

And on 22 April 1943, from the Tiger crews employed in southern Russia: *In general, the observation is made that, when the Panzer is hit by enemy fire, the electrical fuzes melt, causing the vehicle to be temporarily out of service. There are insufficient replacement fuzes. Firing the gun can still continue with the newly delivered inductive emergency firing device; however, the engine can't be restarted. A modification is needed to protect the fuzes so that they will withstand sudden shocks.*

When the turret machinegun is fired without a bag to catch the empty cartridges, the ejected cartridges fall and hit the main battery switch, causing the entire vehicle to suddenly lose electrical power. Therefore it is necessary to install a guard over this switch.

3.4.2.7 Verstarkung der Walzenblende (Reinforced Gun Mantlet)

On 20 March 1942, it was decided that the armor surrounding the sight holes in the gun mantlet would be thicker. The resulting increased size of the outer boring for the gun sight holes had to be accepted. The armor thickness on the right side couldn't be increased, because Wa Pruef 6 was already concerned that the machinegun wouldn't function smoothly through the current penetration. The machinegun itself couldn't be mounted farther out of the turret interior.

On 20 April 1942, D.H.V. stated that, **Verstarkung der Walzenblende** (reinforced gun mantlet) in accordance with change F to drawing 021B863-11 would be implemented with their 41st gun mantlet (**Nr. 250041**). Reinforced gun mantlets were present on some Tigers leaving the assembly plant starting in November 1942.

This change was made too late for Krupp, Essen, which had already completed the 100 gun mantlet castings needed for the **VK 45.01(P)**. Therefore, after **VK 45.01(P)** production was cancelled, starting in January 1943, 90 Tiger I were completed with gun mantlets originally produced for **VK 45.01(P)** turrets which did not have the thicker casting for the gun sight penetrations.

3.4.2.8 Notausstiegluke (Emergency Escape Hatch)

A **Notausstiegluke** was cut into the right rear of the turret side in place of an **M.P.-Klappe** (pistol port) (starting with **Turm Nr.46** in December 1942). The hatch lid was hinged at the bottom and held in place by a vertical bar secured by two hand-screws on the inside of the turret. When the hand-screws were loosened, the bar slid down and released the hatch lid so that it fell open.

Already in January 1943 the **2.Kp./s.H.Pz.Abt.503** reported: *The Notausstiegluke must open the same as a door and be designed with the same type of recessed hinges as the loader's hatch. Currently, the escape hatch can be opened but it can't be closed from inside the turret. The hatch is not only there for escape, it is also used to remove wounded crew members, for communication with infantry, to throw out spent casings, to put out engine fires while in combat by turning the turret to 3 o'clock and spraying with a fire extinguisher out the open hatch, and to aid in recovering broken-down Panzers in combat.*

However, Wa Pruef 6 did not elect to change the hinges on the escape hatch, which remained hinged at the bottom to the end of the Tiger I production series. At first the sides of the lid were beveled to match the curvature of the turret sides, but then uncut flat discs were used for hatch lids starting in July 1943. After a turret ring guard was added in February 1944, the bottom hinge for the escape hatch, bolted to the outside of the turret, was shortened.

3.4.2.9 Verstellbarer Kommandantensitz (Adjustable Commander's Seat)

The commander's seat was modified so that it could be folded up with the bottom of the seat to serve as a back cushion when the commander stood on the turret platform (starting with **Turm Nr.50** in December 1942).

3.4.2.10 Internal Stowage Arrangement

On 19 August 1942, Wa Pruef 6 directed: *The mountings for stowed items may not be welded directly to the inner turret walls, because when the turret is hit the parts will be torn off and endanger the crew. Mountings are to be fastened to the roof or to the turret ring. Electrical connections are to be concentrated onto special sheet metal plates and mounted the same way.* This problem was confirmed by a report from southern Russia in April 1943: *When hit, especially in the turret, all mountings with equipment fly about and wound the crew.*

Additional fasteners and sheet metal holders were installed (starting with **Turm Nr.56** in December 1942). The internal stowage layout was changed again when the new turret design was introduced (starting with **Turm Nr.392** in July 1943).

On 4 May 1944, **Wa J Rue (WuG 6)** informed Henschel, Wegmann, and Krupp: *Daimler-Benz has completed the final design for the **Einheitshalter fuer Scherenfernrohr** for **Pz.Kpfw. und Befehlswagen Tiger I und Tiger II**. The head was changed because it is to be used only for mounting the **Scherenfernrohr** instead of both the **Sehstab** and the **Scherenfernrohr**. The **Einheitshalter** is to be welded underneath the turret roof between the forward periscope and next periscope to the right. The holder for stowing the **Scherenfernrohr** is to be welded under the turret roof to the left front of the Panzer commander. The **Einheitshalter** is to be available shortly from the firm Collignon, Berlin, which is involved in its design.*

3.4.2.11 Minenabwurfvorrichtung (S-Mine Dischargers)

Five **S-Minenwerfer** (S-Mine dischargers) were mounted around the periphery of the hull roof, one at each corner and one halfway along the left hull side. At first only the base mounts were welded to the hull roof (starting in December 1942) until **S-Minenwerfer** became available at the assembly plant (starting by **Fgst.Nr.280082** in early January 1943). Only four **S-Minenwerfer** were mounted on **Panzerbefehlswagen**, one on each corner. By pressing the respective button on the control panel, the **S-Mine** could be selectively fired one at a time.

A separate electrical connection box for the **Minenabwurfvorrichtung** was installed (starting with **Turm Nr.324** in June

1943). As plans were made to install a traversable grenade launcher, fired from inside the turret, the **Minenabwurfvorrichtung** was dropped (in early October 1943).

3.4.2.12 Turmschwenkwerk (Hydraulic Turret Traverse)

On 19 August 1942, Wa Pruef 6 reported: *The design for the foot pedal control for the hydraulic turret traverse has been modified by Krupp and drawings already sent to Wegmann. Approval for the modification will occur following completion of trials by Krupp in Essen. At the same time it was requested that the foot plate for traverse control be made in two pieces that pivot beside each other.*

The drawings for the two-piece foot pedals for the hydraulic traverse drive control were sent to Wegmann at the end of November 1942. This modification, which allowed the gunner to traverse left by pushing down on the left pedal and right with the right pedal, was introduced in Tigers completed in February 1943.

3.4.2.13 Turmbefestigung (Turret Retaining Bolts)

As reported from southern Russia in April 1943: *A large number of bolts holding the turret ring break when the turret is hit by enemy fire. These must be made out of better steel.* Wa Pruef 6 had already introduced a modification to alleviate this problem by increasing the diameter of the bolts holding the turret and cupola and making them out of a better grade of steel (starting with **Turm Nr.160** in March 1943). The number of bolts holding the turret was doubled (starting with **Turm Nr.201** in April 1943).

3.4.2.14 Back Flash Shield

Problems with back flash occurred due to adapting an **8.8 cm Flak** gun and its ammunition as a tank gun inside a closed turret. A sheet metal shield was installed to protect the commander from back flash after the breech opened when firing the **8.8 cm Kw.K.** (starting with **Turm Nr.179** in March 1943). This modification had already been adopted as a backfit to earlier production Tigers, with the following problem noted by the **1.Kp./s.H.Pz.Abt.502** in February 1943: *The sheet metal shield installed to protect the commander against back flashes fulfills this purpose, but it interferes with direct communication between the commander and the loader.* (As in most German Panzers, to allow freedom of movement, the loader was not hooked into the intercom with a pair of earphones.) The proposed solution was to replace the sheet metal shield with a fire-resistant curtain that could be pushed aside. A fire-resistant sailcloth curtain for the commander was installed in place of the sheet metal shield in the new turret design (starting with **Turm Nr.392** in July 1943).

3.4.2.15 Winkelspiegel (Loader's Periscope)

Instead of two periscopes as originally planned in April 1942, a single fixed periscope for the loader was installed facing forward in the turret roof (starting with **Turm Nr.184** in March 1943). Mounted in front of the loader's hatch, the protruding periscope head was protected on the top and sides by an armor guard welded to the turret roof. The two-piece periscope was held in its frame by two wing nuts which pressed the housing into a rubber seal to prevent rainwater intrusion. To cushion the loader's head, pads were fastened to the turret roof and to the periscope housing.

3.4.2.16 Kettengliederhalterungen (Track Holders on Turre

On 19 May 1942, Henschel informed Krupp: *Replaceme track links have to be hung on the turret sides because no oth space is available.* Originally, 15 spare track links were to b stowed around the outside of the turret, but the baggage bin c the turret rear replaced 6 of them.

On 13 April 1943, Henschel informed Wegmann: *The last tv spare track holders on the right side have to be dropped ar another holder mounted farther forward. The five holders on tr left side can be retained. After this change is made, it will allc the engine hatch to be opened when the turret is traversed 1:30.*

Three spare track links were mounted between the visic port and the escape hatch on the right turret side and five spar track links between the vision slit and the pistol port on the le side (starting in mid-April 1943). The number of spare track holc ers on the right side was reduced to two (starting late in Ap 1943).

3.4.2.17 Wiegenzurrung (Internal Travel Lock)

The troops complained that it took about a minute to brir the **8.8 cm Kw.K.36** gun into action because of the time it took t release the internal travel lock. A modified internal travel lock tha held the gun at 0° elevation was installed (starting with **Turr Nr.201** in April 1943).

A new internal travel lock holding the gun at 15° elevatio was installed starting with **Turm Nr.250450** on **Fgst.Nr.25069** in December 1943.

3.4.2.18 M.G.-Lager (Turret M.G. Mount)

Following complaints of numerous stoppages when firing th turret-mounted **M.G.34**, a modified machine gun mount cushione by a spring was installed (starting with **Turm Nr.241** in May 1943

3.4.2.19 Seilzug (Cable Pull for Hatch Lids)

Cable pulls for closing both the loader's and commander' hatch lids from the inside were added (starting with **Turm Nr.29** in June 1943).

3.4.2.20 New Turret

Many major changes that would have interrupted productior if introduced singly were introduced together in a new turret de sign. The exterior shape remained the same, but practically ev ery component was changed (starting with **Turm Nr.392** mountec on **Fgst.Nr.250391** in July 1943), including:
• A **Prismenspiegelkuppel** (commander's cupola with periscopes 021St2761 with a **Ring fuer Flieger-M.G.** (ring mount for an anti aircraft machinegun)
• An **MP-Stopfen** (pistol port plug) 021St48031 in place of the massive pistol port on the left rear of the turret
• A new loader's periscope design, repositioned with a wider ar mor guard
• A modified lock in the center of the **Lukendeckel** (loader's hatch so that the four dogs securing the hatch lid could be loosenec and tightened from the outside. The center shaft was reinforcec initially with an outer disc, which was dropped by October 1943
• A redesigned armor guard (25 mm thick and 300 mm in diam

Chapter 3: Panzerkampfwagen Tiger Ausf.E

er) for the ventilation fan in the turret roof along with relocation the fan to a position as close as possible to directly above the breech of the **8.8 cm Kw.K.** without interfering with the gun travel lock
- A flat disc for the **Notausstiegluckendeckel** (escape hatch lid) without beveled sides
- Addition of a hand control for hydraulic power traverse
- A **Zahnsegment-Turmzurrung** (toothed segment traverse travel lock) replaced the conical pin type
- A new **Federausgleicher** (counterbalance) for the **8.8 cm Kw.K.**, mounted in the turret rear and fastened to the recoil guard by a roller chain, replaced the counterbalance mounted forward along the right turret wall
- A reinforced **Bruecke** (bridge) 021B2861U3 with twice the number of bolts securing it to the turret ring
- A new ball-bearing race with 113 support bearings of 40 mm diameter and 113 separator rings of 55 mm diameter
- A modified sheet metal guard to cover the turret ring gears
- A fire-retardant sail cloth curtain, in place of the sheet metal guard, to protect the commander from back flashes
- The curved armor band, forming the lower turret front, was simplified in shape to make the turret easier to produce
- Other changes to the internal layout of electrical equipment and equipment stowage

The **Prismenspiegelkuppel** (based on 021St48013) and the conical **MP Stopfen** (021St48031) were both borrowed from the turret designed by Krupp for the **8.8 cm Kw.K.43** that was to be mounted on the **VK 45.02(P)** chassis designed by Porsche.

The **Panzer-Fuehrerkuppel (neue Bauart)** (new model of the commander's cupola) consisted mainly of a cast-armor ring, seven periscopes, hatch lid with pivoting arm, and azimuth indicator ring. The cast-armor ring was set into a circular cutout in and welded to the turret roof. Seven evenly spaced openings were cut into the cast-armor ring for mounting seven periscopes in their bakelite housing. The bakelite housings and their flanges were pressed against the rubber seals by tightening two wing nuts. An armor hood was welded over each periscope to protect it against damage from above. The pivoting arm with the hatch cover was released by turning a handwheel that operated a threaded spindle. An operating lever was then used to raise the hatch cover and swing it back to the rear and left. The azimuth indicator ring was carried by six rollers on the inside of the armor ring. It was marked from 1 to 12 o'clock and driven by a shaft connected to a drive gear on the turret ring. This **12-Uhr** azimuth indicator ring in the cupola was dropped from **Pz.Kpfw.Tiger** and **Pz.Bef.Wg.Tiger** starting in early February 1944).

The **MP-Stopfen** (pistol port) in the left turret rear was closed with an armor plug. The armor plug was held in position by a collar that was rotated to the side to release the armor plug. Two chains were attached to the armor plug. The shorter chain held the armor plug after it was pushed out. The longer chain was used to pull the armor plug back into place. **MP-Stopfen** were deleted in October 1943.

Redesigned vision slits (021B2761-44) for both turret sides, with a smaller profile and a wider vision slit to increase the lateral field of view, were introduced in August 1943.

3.4.2.21 Heckzurrung (External Travel Lock)

On 15 December 1942, Wa Pruef 6 approved Krupp's proposed design for an internal travel lock that held the gun at 14° elevation. However, the new internal travel lock wasn't to be introduced into production until an external travel lock to be designed by Henschel became available. A test model of the **Heckzurrung** (external travel lock mounted on the hull rear) was to be completed for testing in Kummersdorf. On 13 April 1943, Wa Pruef 6 asked Henschel to change the design so that the connecting chain was fastened to the other arm and to provide a sheet metal heat guard to protect against heat from the engine exhaust.

A **Heckzurrung** was mounted on the right rear hull (starting with Fgst.Nr.250635 in November 1943). Up to **Turm Nr.250449** on **Fgst.Nr.250692**, the turrets were interchangeable between Tigers. After **Turm Nr.250450** on **Fgst.Nr.250697** with a 15° internal travel lock was introduced, these turrets were exchangeable only with Tigers with a **Heckzurrung**. The **Heckzurrung** was installed only on Tigers with **Fgst.Nr.250635** to **250875** and was no longer mounted (starting with **Fgst.Nr.250876** in February 1944).

3.4.2.22 Turmfugenschutz (Turret Ring Guard)

In their report for the period from 1 to 11 December 1942, the **s.Pz.Abt.501** stated: *During the last action a rare case occurred, in which a clean penetration was made between the edge of the turret and the turret guide ring from the enemy firing from an overlooking position. The turret was jammed and couldn't be repaired until it was first taken out of action for a short period. This was an especially lucky hit that didn't cause any further damage when the projectile entered the fighting compartment. It is proposed that a deflector rail be added to the lower turret edge.*

From another report on the latest experiences with **Pz.Kpfw.Tiger** in Africa dated 24 February 1943: *In two cases the turret was jammed by hits, once by a 3.7 cm projectile and once by artillery shell fragments. A* **Turmfugenschutz** *(turret ring guard) is urgently needed.*

And from the **1.Kp./s.H.Pz.Abt.502** in the northern sector in Russia in February 1943: *In one case a hit in the space below the edge of the turret on Tiger* **Fgst.Nr.250009** *resulted in tearing out eight M16 bolts holding the turret ball race to the superstructure roof and two of the four M12 bolts holding the hand drive for the turret traverse mechanism. In order to prevent this type of heavy damage from hits on the turret, it is proposed that a deflector be added that is set into a ring cut into the superstructure roof.*

On 5 January 1943, Wa Pruef 6 had requested that a deflector rail be designed to protect the turret ring. On 24 February 1943 Krupp reported: *Wa Pruef 6 has specified that the deflector rail withstand hits up to a caliber of 7.5 cm if possible. The deflector rails on the* **Panzerkampfwagen III und IV** *serve only to keep out machinegun bullets and small shell fragments. Other than that the effectiveness of such a rail against 7.5 cm high explosive shells may be overvalued, a significant expenditure would be required to design and attach the rail to the deck. Also, an absolute prerequisite is that the deck be so strong that the rail could be keyed into the deck. It appears to be questionable to us that such a deflector rail can be mounted on the deck. This must be investigated by the chassis designer.*

It wasn't until a year later that Henschel managed to design and introduce a **Turmfugenschutz** (turret ring guard) that was bolted to the superstructure deck at the base of the turret (starting with **Fgst.Nr.250850** in February 1944).

3.4.2.23 Verstaerkung der Turmdecke (Thicker Turret Roof)

In February 1943, **s.Pz.Abt.501** reported on their experience in Africa: *The armor for the turret roof is considered to be too weak because it is feared that large caliber artillery shells detonating on the turret roof will penetrate it. Pz.Kpfw.Tiger lay for a long period in very heavy artillery fire.* In a few cases on the Eastern Front, 15.2 cm shells tore through the 25 mm thick turret roof.

On 8 October 1943, Krupp, responding to a letter from Wegmann dated 21 September 1943, reported that they had initiated the design for strengthening the turret roof. As shown in drawing 2AKF31861 from Krupp dated 30 September 1943, the roof was made out of a single 40 mm thick plate and the commander's cupola redesigned (021B2761U2) for mounting on the thicker roof. A new **Lukendeckel** (loader's hatch lid) without a protective frame (part number 021B50606, borrowed from the Krupp turret designed for the Tiger II) was countersunk into the rear 40 mm thick turret roof plate.

Turrets with thicker 40 mm roof plates were mounted on Tiger I starting with **Fgst.Nr.250991** in March 1944. Originally, these new turret roofs were made as a single 40-mm-thick plate which was bent to slope downward at the front. Starting in May 1944, the turret roofs were constructed using two 40-mm-thick plates which were welded together at the bend directly behind the loader's periscope armor cover. Drain slits were cut into the cast armor ring of the new commander's cupola (021B2761U2), starting in June 1944.

3.4.2.24 Nahverteidigungswaffe (Close Defense Weapon)

A **Nahverteidigungswaffe** was mounted in the roof at the left rear (at the same time as the 40 mm thick turret roof, starting in March 1944). The **Nahverteidigungswaffe** was used to fire the **Sprenggranatpatrone 326 Lp** (explosive shells), **Schnellnebelkerzen 39** (quick smoke candles), **Rauchsichtzeichen orange 160** (orange smoke signals), and **Leuchtgeschossen R** (flare signals) with the **Zundschraube C 43 St** (propellant charge). The **Nahverteidigungswaffe**, aimed at a fixed angle of about 50°, could be traversed 360°. An armor plug could be inserted into the bore to seal it when not in use.

The **Sprenggranatpatrone 326 Lp** with a 1 second delay fuze was propelled out to 7 to 10 meters and exploded about 0.5 to 2 meters above the ground. Individual fragments flew in a circle for up to 100 meters. All turret hatches and openings were to be closed when this round was fired.

3.4.2.25 Monokular Turmzielfernrohr (Monocular Gun Sight)

On 2 February 1943, Wa Pruef 6 alerted Krupp that a monocular gun sight would be introduced shortly into the Panther and Tiger (1 & 3) turrets. Krupp was to propose a suitable armor plug to close the left hole already bored in completed gun mantlets. Starting in late March/early April 1944, the monocular **T.Z.F.9c** replaced the binocular **T.Z.F.9b** telescopic gun sights. The magnification of the **T.Z.F.9c** could be switched between 2.5x (28 degree field of view) and 5x (14 degree field of view).

3.4.2.26 Leichte Muendungsbremse (Lighter Muzzle Brake)

Starting in April 1944, some of the Tigers were outfitted with a **leichte Muendungsbremse** (lighter muzzle brake) on the **8.8 cm Kw.K.36 L/56**. These muzzle brakes designed for the longer **8.8 cm Kw.K.43 L/71** had replaceable ring inserts. On 6 April 1944, Wegmann asked Krupp about drawings with the corresponding changes to the **Federausgleicher** (spring counterbalance) caused by the lighter weight gun.

3.4.2.27 Pilze fuer Behelfskran 2t (Sockets for Jib Boom)

Starting in June 1944, a **Behelfskran 2t** (jib boom) was issued to the troops to aid in tank repairs. This **Behelfskran** was mounted on three **Pilze** (sockets) welded to the turret roof. The **Behelfskran 2t** could be used to lift the rear decking, motor, and suspension components from the Panzer on which it was mounted or to lift components from an adjacent vehicle.

3.4.2.28 Reparierte Tiger H1 Tuerme (Recycled Turret Armor)

The last 54 Tiger I, **Fgst.Nr.251293 – 251346**, were to be produced utilizing 32 turrets recycled from Tigers that had been returned from the front for major rebuild as well as 22 new turrets produced under replacement parts contracts. In a meeting on 1 June 1944, D.H.H.V., Krupp, and Henschel determined the changes that were needed to "modernize" the recycled turrets to meet the latest standards:

*For modernization, the following modifications to the armor are to be accomplished by the **Panzerfirmen** (D.H.H.V. and Krupp):*
- **Decke** 021B2761-21 (turret roof plate) has been increased from 25 to 40 mm thick. Cut off the old 25 mm thick plates and weld on two new 40 mm thick plates.
- **Mantel zur Kommandantenkuppel** 021C862-15 (armor ring for the commander's cupola) The shape has been changed, and the older cupola can't be used on the thicker turret roof plates. Replace with the new cast cupola ring 021B2761-U2 and weld to the turret roof.
- **Mantel** 021B2761-20 (U-shaped turret wall) Holes for the **M.P. Klappe** or **M.P.-Stopfen** have been deleted. Plug the hole for the **M.P.-Klappe** or deliver the **M.P.-Stopfen** loose.
- **Lukendeckel** 021C865-U3 (loader's hatch) shape has changed. Replace the old loader's hatch lid with the new **Lukendeckel** 021B50606-U1 and deliver loose.

All other changes to the turrets will be accomplished by the assembly firm.

3.4.2.29 Losterkennungstafel (Poison Gas Detection Panels)

At a meeting on 22 and 23 June 1944 at Wegmann, Wa Pruef 6 discussed plans to install poison gas protection equipment in the **Pz.Kpfw.Tiger E**: *A filter is to be mounted behind the spring counterbalance and the gas protection box forward to the left of the gunner in the Tiger E turret. The only location available for mounting the **Losterkennungstafel** is on the turret. On the Tiger Ausf.E one panel is to be mounted on the armor sleeve for the gun about 400 mm in front of the gun mantlet, a second panel (bent at about 15°) on the left side of the gun mantle, and a third panel (bent at 25°) about the middle of the baggage bin. It isn't possible to install this poison gas protection equipment during assembly of the **Tiger E** because series production is scheduled to run out in mid-July. Therefore this equipment must be backfitted.*

Assembly of Tiger I turrets at Wegmann actually continued into August 1944, and this modification was added by Wegmann to Tiger II turrets by 9 July 1944. Therefore, poison gas protection equipment may have been installed on the assembly line before the last Tiger I turrets were completed at Wegmann during July and August 1944.

Chapter 3: Panzerkampfwagen Tiger Ausf.E

EXTERNAL MODIFICATIONS INTRODUCED DURING THE TIGER I PRODUCTION RUN

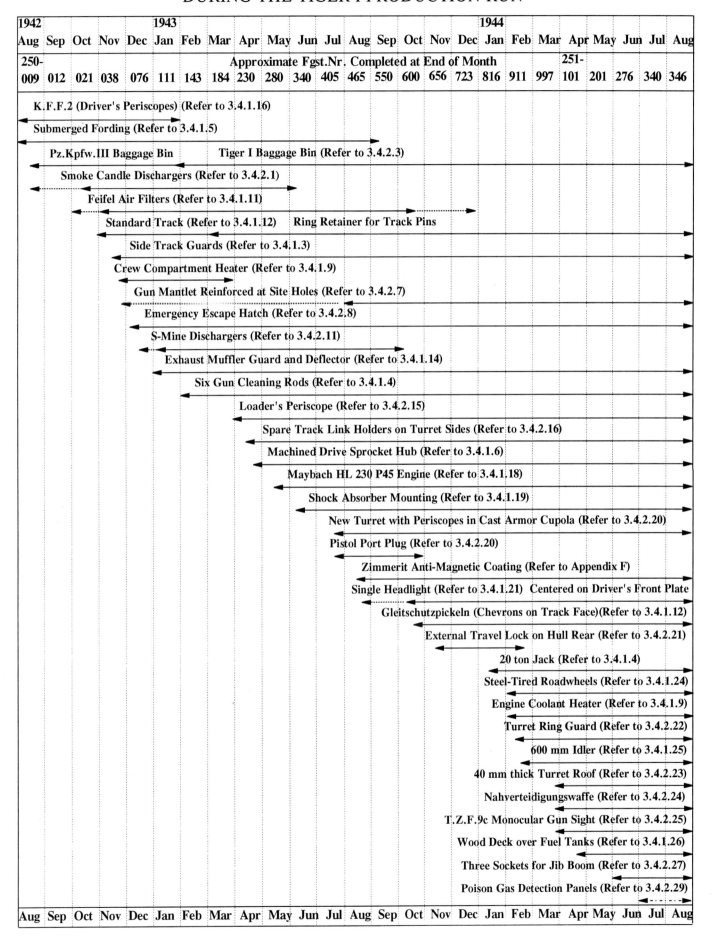

Wa Pruef 6 confiscated the first production series chassis (**Fgst.Nr.250001** completed in May 1942) for testing. It is shown here after the steerin mechanism broke. Periscopes had been installed in the hatch lids for the driver and radio operator but, as was the usual practice with Wa Pruef test vehicles, it was not outfitted with any tool stowage. (BA)

Chapter 3: Panzerkampfwagen Tiger Ausf.E

As shown on the bottom of **Wanne Nr. 25002(3)** on the assembly line in October/November 1942, openings in the bottom of the hull included: a drain cock in the front (vehicle right) by the radio operator's seat, a drain cock at the front (vehicle left) of the engine compartment, a drain cock at the rear (vehicle right) of the engine compartment, a small opening (cover plate missing) toward the front left under the steering gear, a small opening (cover plate missing) under the transmission, and two cover plates under the engine. (TTM)

THIS PAGE AND OPPOSITE: **Pz.Kpfw.VI H Ausf.H1** (**Fgst.Nr.250031** completed in November 1942) in the original condition in which it was reassembled at Aberdeen Proving Ground. As shown by the Tiger stenciled on the driver's front plate and the tactical numbers on the turret, it was captured from **s.Pz.Abt.501** in Tunisia. The Feifel air cleaner on the left rear with markings from the **2.Kp./s.Pz.Abt.504** and **Pz.Abt.215** (G in a star for the commander, Major Gierga) was taken off another Tiger returned to Aberdeen Proving Ground from Sicily. (APG)

Chapter 3: Panzerkampfwagen Tiger Ausf.E

Details of the turret components on **Pz.Kpfw.VI H Ausf.H1** (**Fgst.Nr.250031** completed in November 1942) (TLJ)

The front of the turret roof was reinforced with a 40 mm thick strip. With the **8.8 cm Kw.K.** centered in the turret, it is easy to observe that the turret was extremely asymmetrical, with the left 110 mm wider than the right.

The right side of the gun mantlet, with details of the aperture for the turret machinegun.

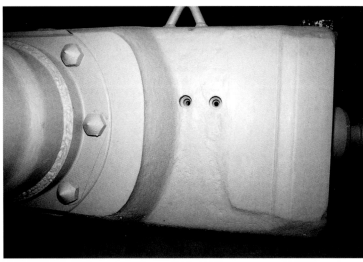

The left side of the gun mantlet, with details of the apertures for the binocular gunsight.

Details of the counterbalance spring and retention stop for the commander's cupola hatch lid.

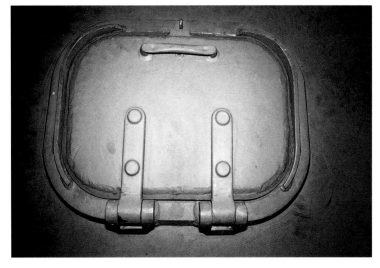

A D.H.H.V. loader's hatch constructed of two plates welded together. All loader's hatches were mounted in a deflector frame on the 25 mm thick turret roofs.

A pistol port was mounted on the right rear of the first 45 turrets completed by Wegmann.

Chapter 3: Panzerkampfwagen Tiger Ausf.E

Details of the exterior of **Pz.Kpfw.VI H Ausf.H1** (**Fgst.Nr.250031** completed in November 1942) (TLJ)

The machinegun ballmount flanked by pivoting toggle bolts for retaining cover seal during submerged fording.

The driver's visor in the open position with apertures for the binocular **K.F.F.2** driver's periscopes. The fastener on the glacis is for holding a shovel.

The reinforcing strip with five bolts to secure the final drive housing. Holes in the hull side extension, originally intended to be used for the pivoting **Vorpanzer** (spaced armor), have been plugged and welded closed. The **Wanne Nr.250058 amp** is stamped into the armor (in the rectangle free of paint). "amp" was the secret three-letter code for D.H.H.V., the hull manufacturer.

The drive sprocket wheel with sprocket teeth and ring bolts positioned out of phase with the hub spokes.

The 700 mm diameter idler with scooped center to aid in clearing out heavy mud.

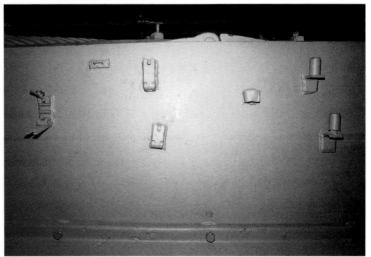

Single and double retainers and palls for holding the track replacement cable on the side of the hull. Smaller wire tabs were used to hold a canvas tarp in place across the engine deck.

Panning clockwise around the turret interior of **Pz.Kpfw.VI H Ausf.H1** (**Fgst.Nr.250031**). Compare these photographs to those from the interior the turret on **Fgst.Nr.250122**. (TLJ)

The internal travel lock above the breech of the **8.8 cm Kw.K.36**.

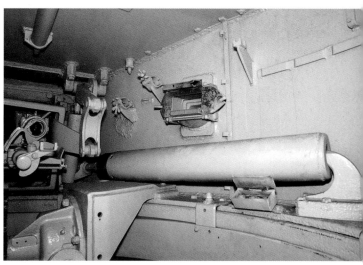

The coaxial machinegun mount, spring counterbalance, and vision bloc on the right front side.

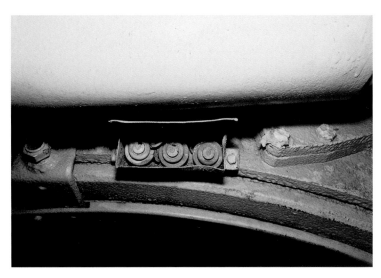

The three buttons used to set off the three smoke candle dischargers on the right side of the turret.

The pivoting slide used to open and close the machinegun port (in the closed position).

Clips to secure an **M.P.38** machine pistol, a box to hold 12 signal pistol cartridges, and frames to store spare glass vision blocks along the turret rear.

At the far right is the upper commander's seat (folded down without its cushion), curved brackets to hold a gas mask, and the pistol port on the left turret rear.

Chapter 3: Panzerkampfwagen Tiger Ausf.E

Details of the hull interior of **Pz.Kpfw.VI H Ausf.H1** (**Fgst.Nr.250031** completed in November 1942). Compare these photographs to those from the hull interior of **Fgst.Nr.250122**. (TLJ)

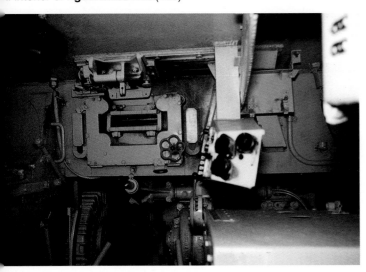

The driver's position to the left with the frame slide for mounting the **K.F.F.2** driver's periscopes above the driver's visor which was opened and closed by the five-spoke handle. The radio racks mounted over the transmission housing are missing.

The radio operator's position with the ball mount for the hull machinegun. The radio racks mounted over the transmission housing are missing.

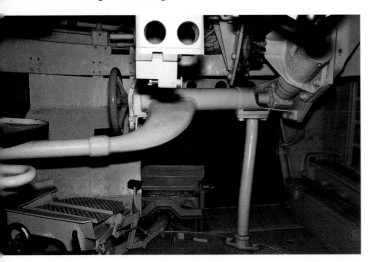

Under the gun looking forward, with the gunner's seat on the left, elevation wheel with main gun firing trigger in the center, the foot pedal for traversing the turret, and the foot trigger for firing the machinegun (linkage disconnected).

Ammunition racks on the right side, with each bin in the sponson holding 16 rounds and each lower bin holding 4 rounds.

The **Fgst.Nr.250031 dkr 42** was stamped into the side support (where the paint is missing) between the ammunition racks on the right hull side. "dkr" was the secret three-letter code for Henschel.

The firewall separating the engine compartment from the crew compartment, with an automatic fire extinguisher mounted on the left and a rectangular cold start fuel injector on the right.

95

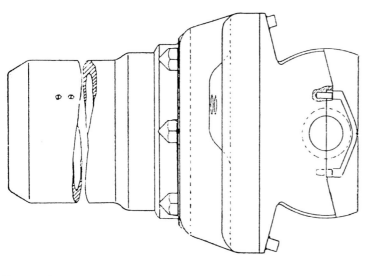

Verstaerkung der Walzenblende (reinforcing the gun mantlet) at the gunsight apertures was introduced by D.H.H.V. with their 41st gun mantlet (**Nr. 250041**).

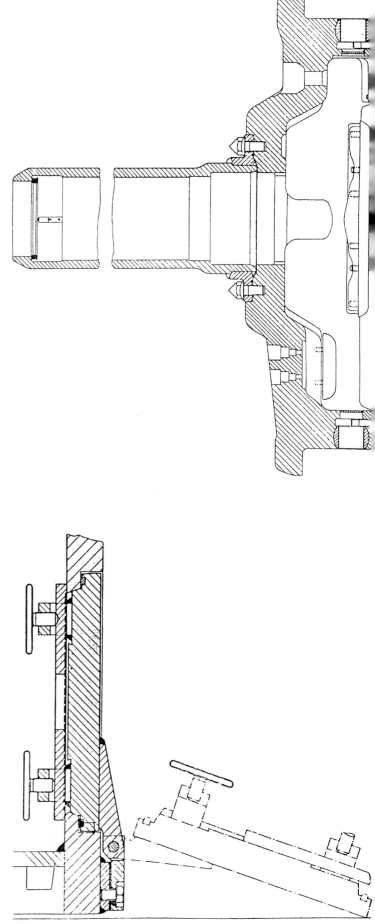

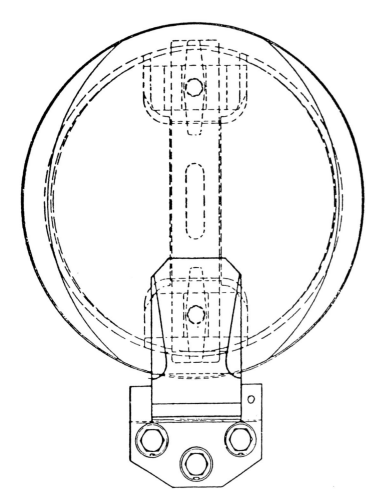

A **Notausstiegluke** was cut into the right rear of the turret side in place of an **M.P.-Klappe** (pistol port) (starting with **Turm Nr.46** in December 1942).

Chapter 3: Panzerkampfwagen Tiger Ausf.E

A propaganda photograph was taken of Japanese officers in the turret of **Pz.Kpfw.VI H Ausf.H1** (**Fgst.Nr.250055** completed in December 1942). New features included redesigned hull side extensions (to allow the tow shackle to swing free), a reinforced gun mantlet at the gunsight apertures, and leveling cylindrical bases for a tubular frame to support a camouflage canopy. (TTM)

Completed in December 1942 and issued to the **2.Kp./s.Pz.Abt.502** (renamed **3.Kp./s.Pz.Abt.503**), this is one of the first **Pz.Kpfw.VI H Ausf.H1** with the emergency escape hatch in the turret in place of the machinegun port. The armor guards over the penetration in the hull roof for the headlight wiring had been redesigned. It still has normal stowage with five gun cleaning rods but has been modified by the unit (welding the driver's periscope holes shut, mounting a larger stowage box on turret rear, mounting a fire extinguisher beside the stowage box, and adding brackets to hold water cans on the rear). (BA)

 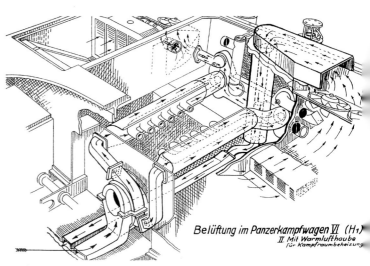

Starting in December 1942, **Pz.Kpfw.VI H Ausf.H1** sent to Russia were outfitted with a **Kampfraumbeheizung** (crew compartment heater) consisting of a sheet metal cowling that redirected air warmed by the left radiator into the top of the air intake snorkel. (NA)

The ventilation flow path when the **Warmlufthaube** (warm air cowling) was installed for crew compartment heating.

This Tiger, completed in January 1943 and issued to **s.Kp./Pz.Rgt.GD**, has five **S-Mine** dischargers, standard sheet metal guards for the mufflers and an oversized stowage bin (in place of the factory supplied **Pz.Kpfw.III** stowage bin). (WS)

Chapter 3: Panzerkampfwagen Tiger Ausf.E

The third **Pz.Kpfw.VI H Ausf.H1** (**Fgst.Nr.V3**) of the **Versuchs-Serie** was completed in January 1943 with **Turm Nr.88** with an emergency escape hatch (instead of the turret ordered for this trial series from Krupp). The antenna base is still on the hull rear, but it was completed with the latest style of track guards. The cap sealing the machinegun ball mount is in place and the snorkel raised on the rear deck for conducting submerged fording experiments. (WJS)

Panzerkampfwagen VI H Ausf.H1 – Fgst.Nr.250122 completed in February 1943 – extended hull sides – plugged apertures for **K.F.F.2** driver periscopes – leveling ring above holes for camouflage frame – hinged track guards front and rear with straight side track guards – mounts for S **Mine** dischargers – six short gun cleaning rods – Feifel air cleaners mounted – Krupp gun mantlet originally cast and cut out for a Tiger (P) emergency escape hatch – tubular sockets on cupola as bases for rain shield

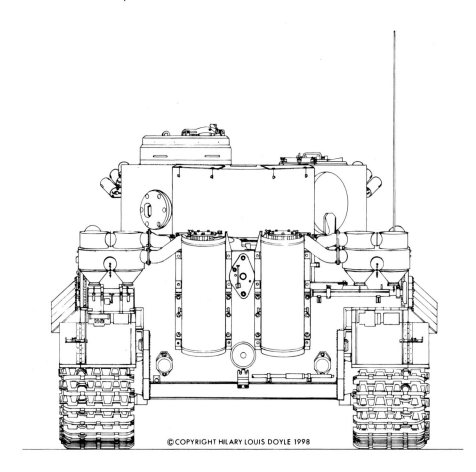

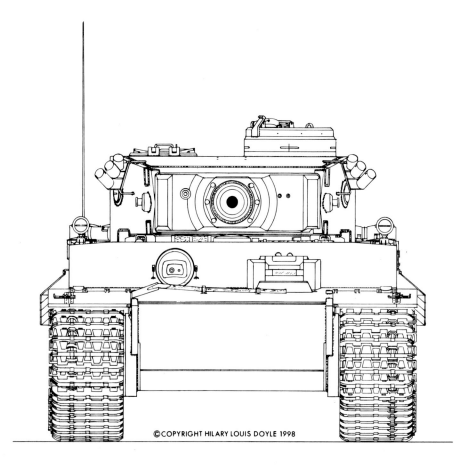

Chapter 3: Panzerkampfwagen Tiger Ausf.E

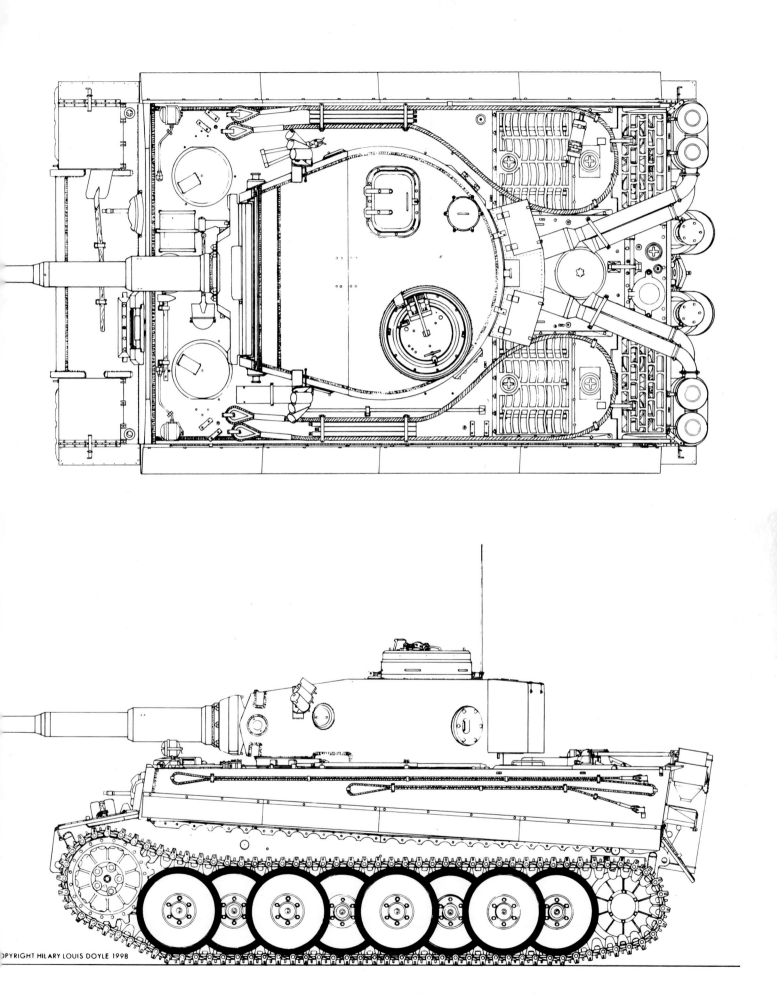

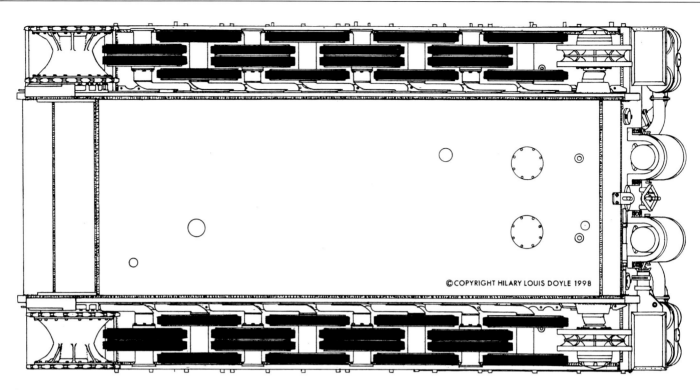

Panzerkampfwagen VI H Ausf.H1 – Fgst.Nr.250122 completed in February 1943 – dropped forward drain valve – outer roadwheels and track guards removed for rail loading

Pz.Kpfw.VI H Ausf.H1 (Fgst.Nr.250122 completed in February 1943 and issued to the **1.Kp./s.Pz.Abt.504**) before it was captured in Tunisia. This is the Tiger I currently being restored at the Tank Museum in Bovington, England. (BA)

Chapter 3: Panzerkampfwagen Tiger Ausf.E

Pz.Kpfw.VI H Ausf.H1 (**Fgst.Nr.250122** completed in February 1943) with **Marschketten** (725 mm wide operational tracks) and **Verladeketten** (520 mm wide transport tracks). In the right-side view, the outer roadwheels have been removed revealing the roadwheel rims reinforced by two bolts flanking each of the original six bolts. (TTM)

The forward tow cable fasteners were moved back and the spare antenna rod stowage tube moved forward in December 1942 to make room for S-Mine discharger mounts. The headlights were removed, the driver's hatch was replaced by a radio operator's hatch, and the shattered loader's hatch was replaced after **Pz.Kpfw.VI H Ausf.H1** (**Fgst.Nr.250122**) was captured by the British. (TTM)

The interior of the engine compartment directly below the engine deck hatch revealing the air intake manifold for the **Maybach HL 210 P45** engine with three air filters (TTM)

Chapter 3: Panzerkampfwagen Tiger Ausf.E

Panning clockwise around the turret interior of **Pz.Kpfw.VI H Ausf.H1 (Fgst.Nr.250122)** (TTM)

The breech of **8.8 cm Kw.K.36 R 179 42 amp** with the coaxial **M.G.34** mounted to the right

The spring counterbalance for the gun was mounted on the right turret ring

Stowage along the right turret wall to the emergency escape hatch

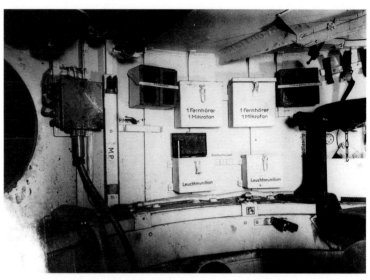

Stowage along the rear turret wall to the commander's seat

Stowage along the left turret wall including instructions for closing the seals in the turret for submerged fording

Stowage in the left front corner included the gunner's azimuth indicator (synchronized to match the azimuth indicator ring in the commander's cupola).

Panning clockwise around the hull interior of **Pz.Kpfw.VI H Ausf.H1 (Fgst.Nr.250122) (TTM)**

Ammunition stowage to the left of the driver with a compass mounted in the corner

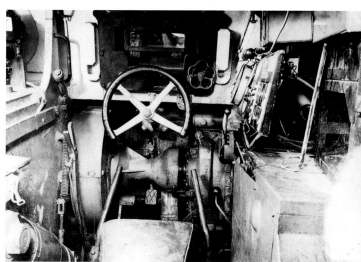

The driver's position with a Henschel **Osilit** steering wheel with four spokes

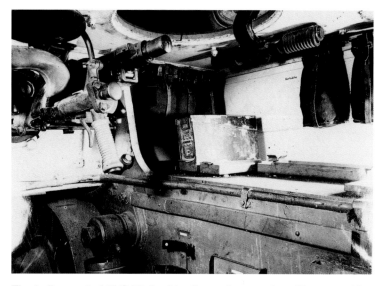

The ball-mounted **M.G.34** fired by the radio operator with ammunition stowage in the pannier to the right

The **Fgst.Nr.250122 dkr 43** was stamped into the side support between the ammunition racks on the right hull side

The access hatch lifted to reveal the ammunition stowed on the right side under the turret platform

The bilge pump discharge pipe, fire extinguisher and cold start fuel injector along the firewall behind the brackets for holding three water cans and the wire mesh basket for signal flags

Chapter 3: Panzerkampfwagen Tiger Ausf.E

The driver's position in a **Pz.Kpfw.VI H Ausf.H1** completed in February 1943 with an **Argus-Lenkapparat L.St.0.2** (steering device). Holes for the **K.F.F.2** driver's periscopes have been plugged. (GF)

Pz.Kpfw.Tiger Ausf.E (Fgst.Nr.250159 completed in March 1943) with ring retainers for the track pins. The unit has sprayed camouflage paint onto the gun barrel which would burn and turn black after rapid firing. (MJ)

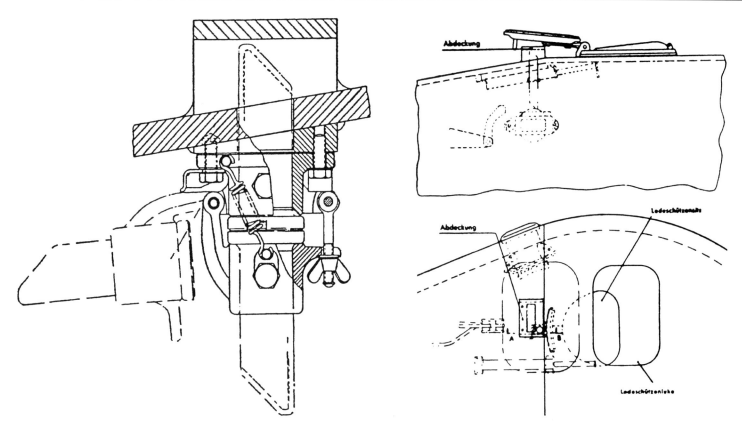

A single **Winkelspiegel** periscope for the loader was installed facing forward in the turret roof (starting with **Turm Nr.184** in March 1943).

When this modification was introduced in mid-April 1943, Wegmann welded four track holders on the right turret side and five on the left turret side. The rear two on the right side were dropped after Henschel discovered that the engine compartment hatch couldn't be opened. (NA)

Chapter 3: Panzerkampfwagen Tiger Ausf.E

Pz.Kpfw.Tiger Ausf.E (within **Fgst.Nr.** series **250200** to **250208** completed in April 1943 and issued to the **2.Kp./s.Pz.Abt.504**) with three track holders on the right turret side and single upper chamber **Feifel** air cleaners. (NA)

A closeup view of the simplified **Feifel** air cleaner with a single upper chamber design (introduced in March 1943). (BA)

Chapter 3: Panzerkampfwagen Tiger Ausf.E

IMAGES 111-113: Four views of **Pz.Kpfw.Tiger Ausf.E** (**Fgst.Nr.250234**, the fourth completed in May 1943) with a loader's periscope on the turret roof, spare track link holders on the turret sides, a machined drive sprocket wheel hub, and 12 bolts retaining the steel rim on each roadwheel. It has D.H.H.V. turret with welded loader's hatch and reinforced gun mantlet by the gun sight apertures. These are propaganda photos taken at the Henschel assembly plant. Just like all other Panzers, Tigers were outfitted at a **Heeres Zeugamt** (ordnance depot) before being released to a unit. (LD)

Chapter 3: Panzerkampfwagen Tiger Ausf.E

GERMANY'S TIGER TANKS - D.W. to TIGER I

Four views of **Pz.Kpfw.Tiger Ausf.E** (**Fgst.Nr.250235**, the fifth completed in May 1943, issued to the **3.Kp./s.Pz.Abt.502**) with a machined h drive sprocket wheel on the left side but the older style hub drilled for each mounting bolt on the right side. It has a Krupp turret with forged loade hatch and a gun mantlet cut out on the lower left corner to clear the air louvers on a Tiger (P). (BA)

Chapter 3: Panzerkampfwagen Tiger Ausf.E

The **Pz.Kpfw.Tiger Ausf.E** in the foreground with three spare track link holders on the right turret side has already received a base coat of **Dunkelgelb** (RAL 7028) paint, while **Fgst.Nr.250238** (the 8th completed in May 1943 just leaving the assembly hall for acceptance tests) is still coated with the red oxide primer (RAL 8012). (HLD)

Chapter 3: Panzerkampfwagen Tiger Ausf.E

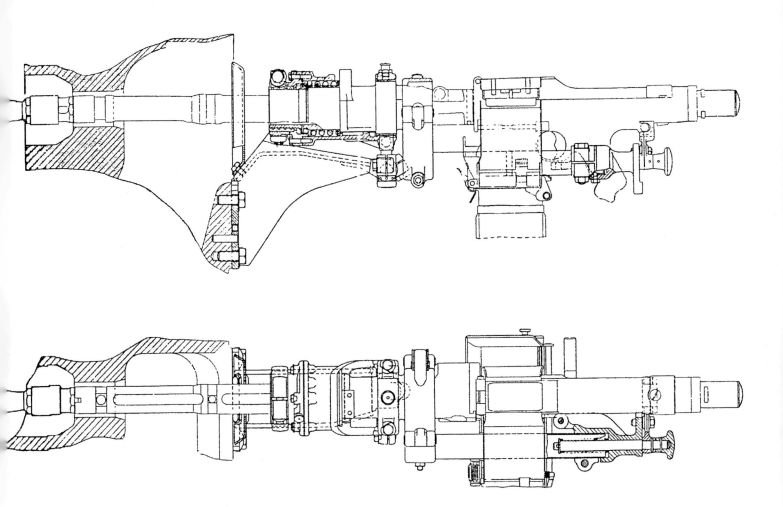

A modified machinegun mount for the coaxial **M.G.34**, cushioned by a spring, was installed (starting with **Turm Nr.241** in May 1943).

Starting with **Fgst.Nr.250251**, the larger Maybach **HL 230 P45** engines were installed in **Pz.Kpfw.Tiger Ausf.E**, which is externally identifiable by two holes in the crank starter base plate. At the same time, the frames welded to the hull rear for mounting the rear fenders and tail lights were dropped. They were replaced by new hinged rear track guards (with cut-outs for the convoy light), supported by three hinge-plates welded directly to the hull rear (which have been torn off of this Tiger). (BA)

Chapter 3: Panzerkampfwagen Tiger Ausf.E

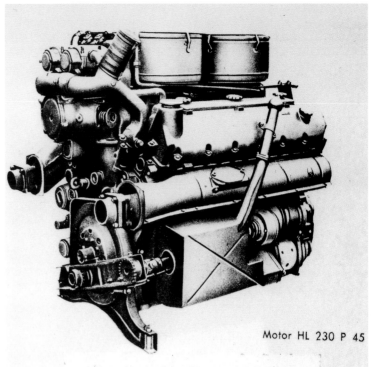

The 23 liter **HL 230 P45** engine with both magnetos mounted behind the twin air filters. (NA)

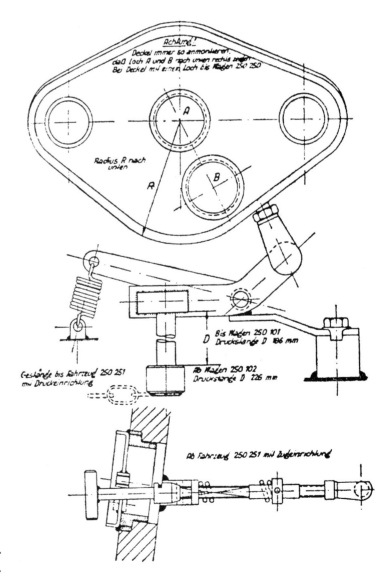

A second hole was added to the **Deckel fuer Schwungkraftanlasser** (crank starter base plate) to align the starter shaft with the Maybach **HL 230 P45** engines.

GERMANY'S TIGER TANKS - D.W. to TIGER I

Panzerkampfwagen Tiger Ausf.E – Fgst.Nr.250251 completed in May 1943 – drive sprocket hub machined flat – ring retainer for track pins – S **Mine** dischargers mounted – "triangular" access port on rear deck – same casting on both sides for rear louvers – modified Feifel air cleaners dropped frames on hull rear for mounting track guards – second hole in starter alignment plate for **HL 230 P45** engine – gun mantlet reinforced b gunsight apertures – loader's periscope – spare track holders on turret sides

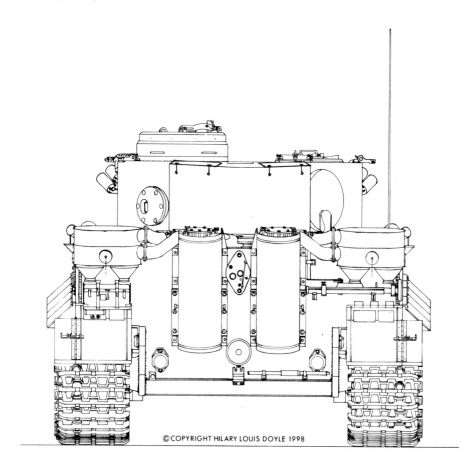

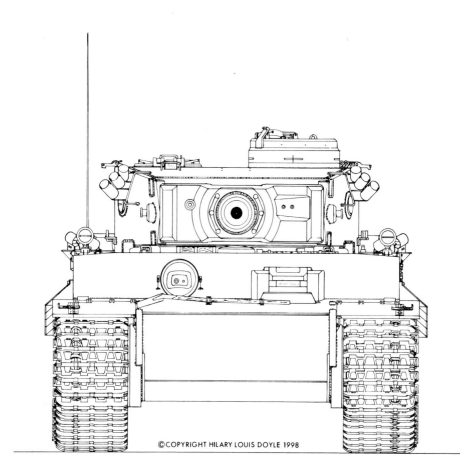

Chapter 3: Panzerkampfwagen Tiger Ausf.E

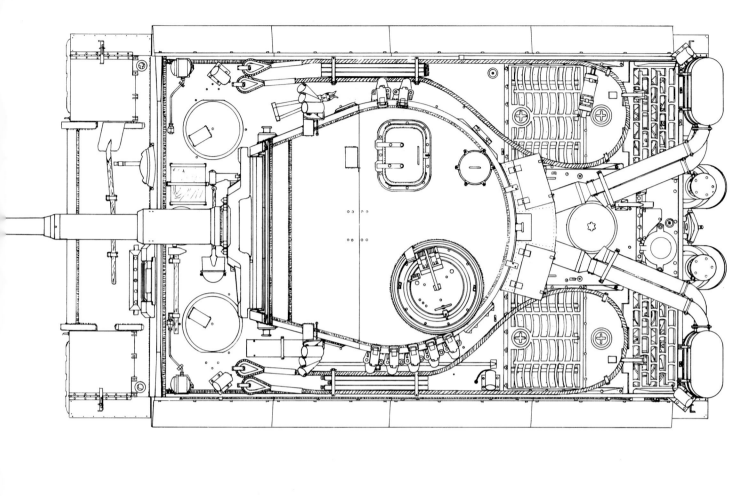

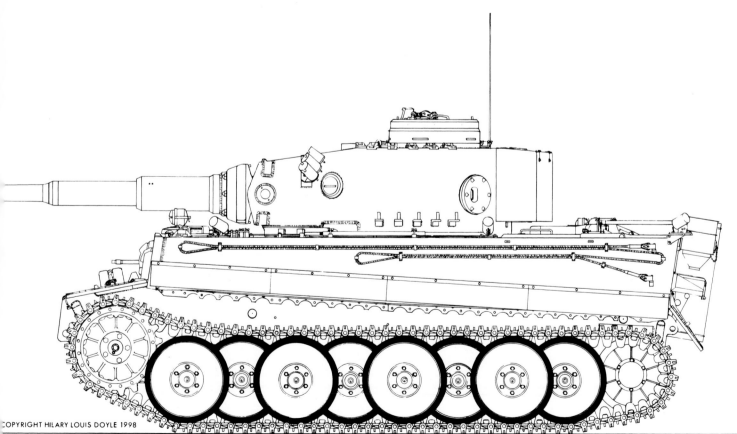

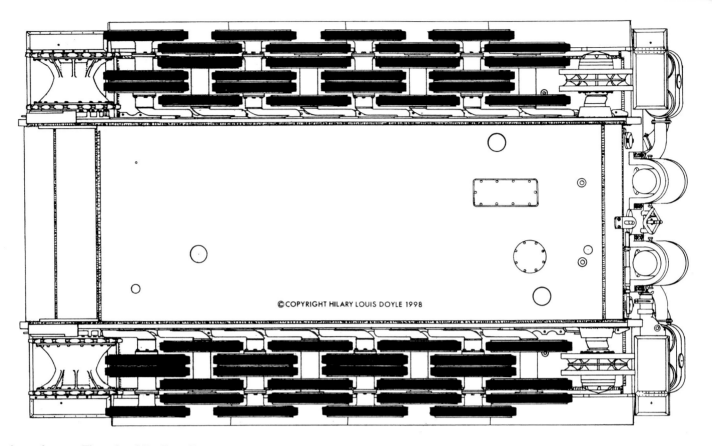

Panzerkampfwagen Tiger Ausf.E – Fgst.Nr.250251 completed in May 1943 – large rectangular hole for access to the electrical generator and fuel pumps – added another fuel drain port

The smoke candle dischargers were dropped (starting with **Turm Nr.286**) and the toggle bolts to hold the sealing cap onto the machinegun ball mount were dropped in June 1943 as shown on this **Pz.Kpfw.Tiger Ausf.E** completed in July 1943. (BA)

Chapter 3: Panzerkampfwagen Tiger Ausf.E

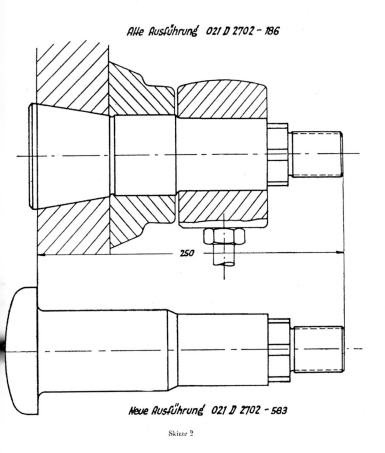

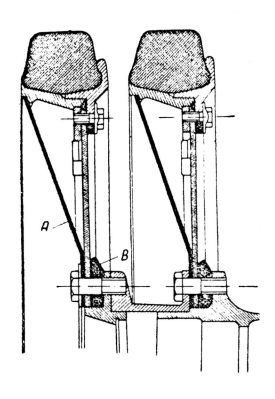

The mounting bolt for shock absorbers was changed to one with a large external head (starting with **Fgst.Nr.250301** in June 1943).

The roadwheels were strengthened at the base of the discs by being fastened with bolts and reinforced by a weld bead (starting with **Fgst.Nr.250314** in June 1943).

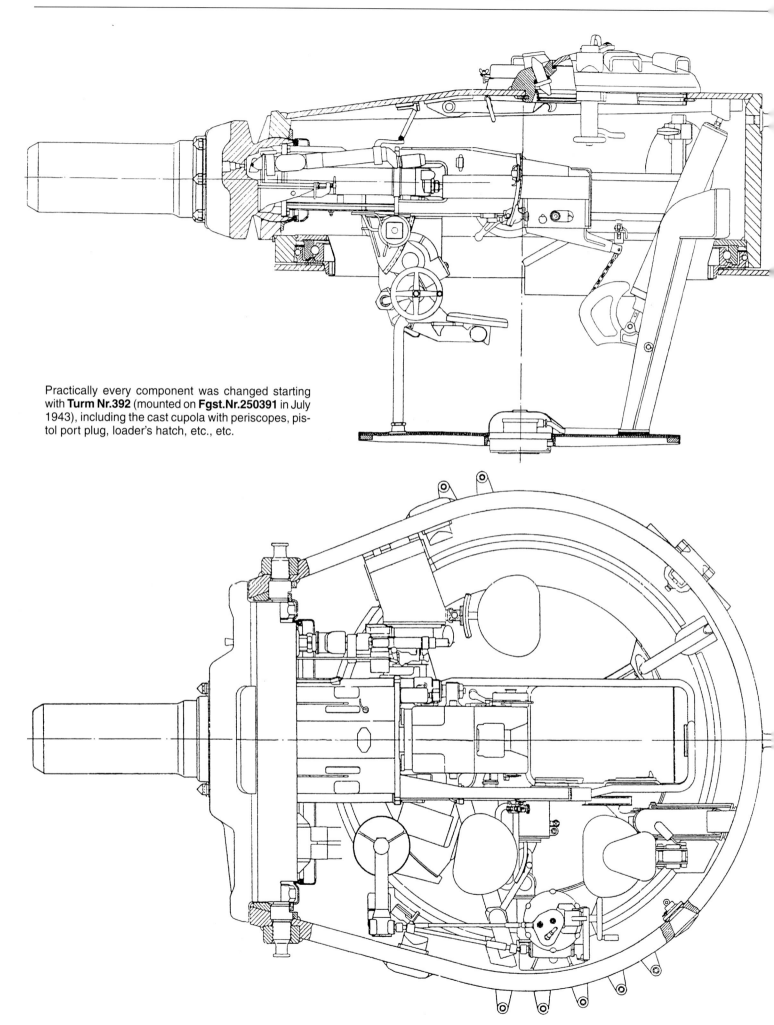

Practically every component was changed starting with **Turm Nr.392** (mounted on **Fgst.Nr.250391** in July 1943), including the cast cupola with periscopes, pistol port plug, loader's hatch, etc., etc.

Chapter 3: Panzerkampfwagen Tiger Ausf.E

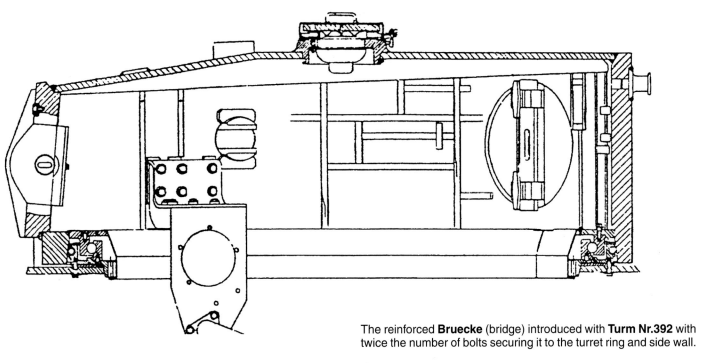

The reinforced **Bruecke** (bridge) introduced with **Turm Nr.392** with twice the number of bolts securing it to the turret ring and side wall.

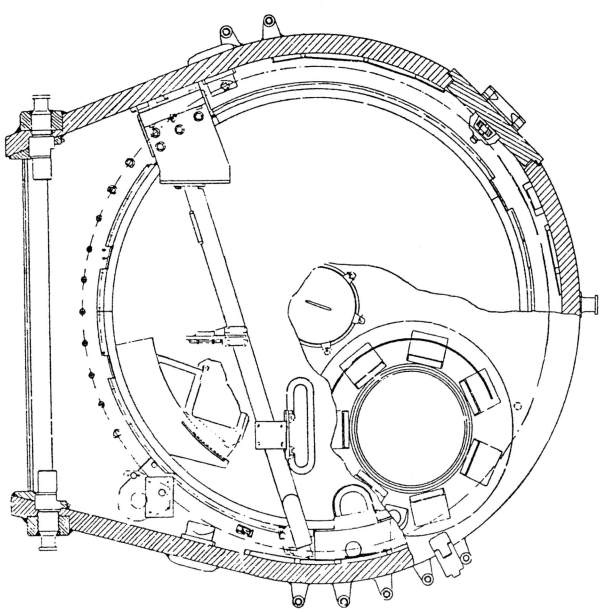

The cast armor **Prismenspiegelkuppel** (commander's cupola with periscopes introduced with **Turm Nr.392**) with a **Ring fuer Flieger-M.G.** (ring mount for an anti-aircraft machinegun) welded to the periscope guards and a 12-hour azimuth indicator ring mounted inside.

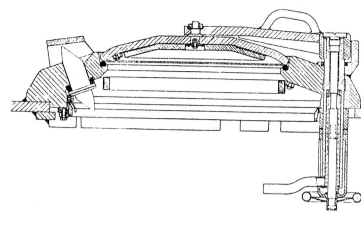

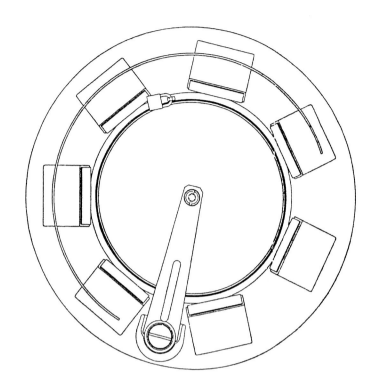

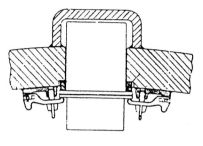

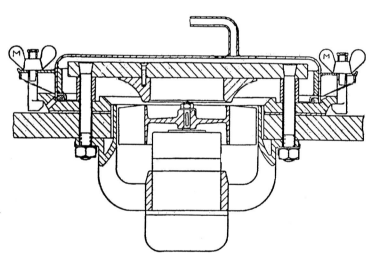

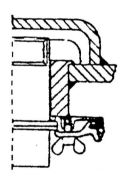

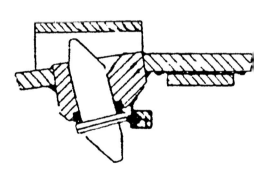

The mounting for the loader's periscope was redesigned to include a wider armor cover (starting with **Turm Nr.392** in July 1943).

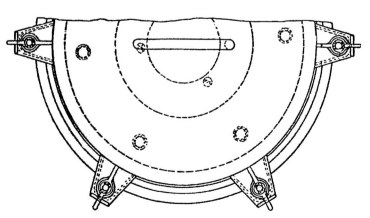

The ventilation fan was relocated to a position close above the breech of the **8.8 cm Kw.K.** and the design of the armor guard and sealing cap changed (starting with **Turm Nr.392** in July 1943).

Chapter 3: Panzerkampfwagen Tiger Ausf.E

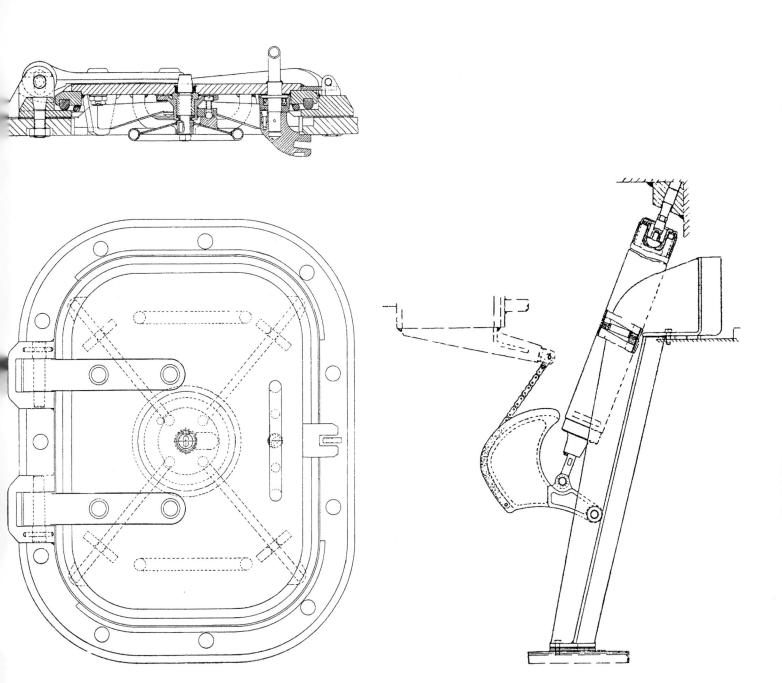

A modified lock was added to the center of the **Lukendeckel** (loader's hatch) so that the four dogs securing the hatch lid could be loosened and tightened from the outside (starting with **Turm Nr.392** in July 1943).

A new **Federausgleicher** (spring counterbalance) for the **8.8 cm Kw.K.36**, mounted in the turret rear and fastened to the recoil guard by a roller chain, replaced the counterbalance mounted forward along the right turret wall (starting with **Turm Nr.392** in July 1943).

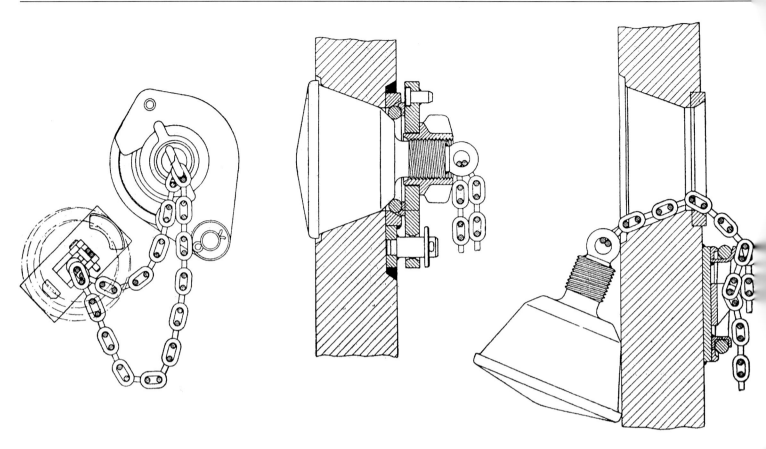

MP-Stopfen (pistol port plug) replaced the **MP-Klappe** (starting with **Turm Nr.392** in July 1943). Two chains were fastened to the plug: the shorter chain to hold the ejected plug, the longer chain to pull the plug back into place.

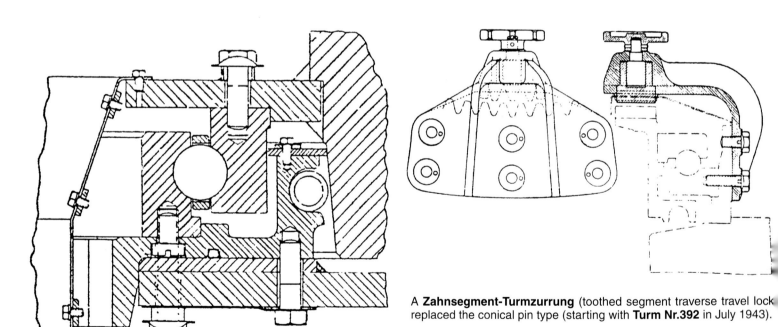

A **Zahnsegment-Turmzurrung** (toothed segment traverse travel lock) replaced the conical pin type (starting with **Turm Nr.392** in July 1943).

A new **Turmkugellager** (ball-bearing race) with 113 support bearings of 40 mm diameter and 113 separator rings of 55 mm diameter was introduced (starting with **Turm Nr.392** in July 1943). A modified sheet metal guard covered the turret ring gears.

Chapter 3: Panzerkampfwagen Tiger Ausf.E

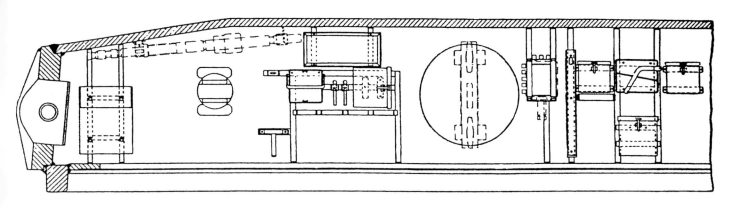

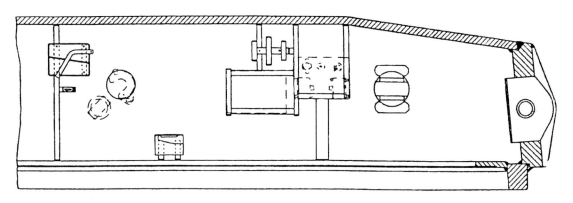

Stowage inside the turret was altered (starting with **Turm Nr.392** in July 1943) mainly due to the modification of the bridge, spring counterbalance, and cupola. Compare these drawings to photographs of the stowage layout inside the turret on **Fgst.Nr.250122**.

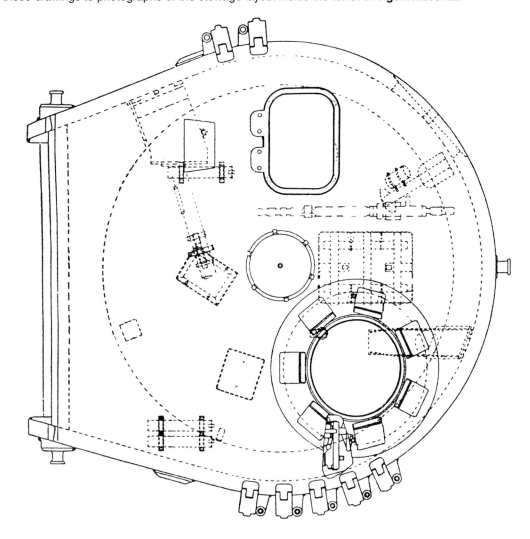

THIS PAGE AND OPPOSITE: **Pz.Kpfw.Tiger Ausf.E** (**Fgst.Nr.250427** completed in August 1943) on display in Kubinka, Russia is one of the **Panzerbefehlswagen** converted back to normal **Panzerkampfwagen** in November 1943. The large base (for protecting a porcelain insulator the **Sternantenne**) has been plated over and the penetration for the antenna base on the turret roof plugged. The exhaust port from the **GG4** generator set was located on the deck next to the mounts for supporting extension rods for an elevated **Sternantenne**. The antenna base on the l side (normally used for long range communication with aircraft by the **Fu 7** radio set in a **Panzerbefehlswagen**) was used for the short range **Fu** radio set in this conversion. (Please, disregard the fake rear fenders, muffler caps, and Russian headlights) (TA)

Chapter 3: Panzerkampfwagen Tiger Ausf.E

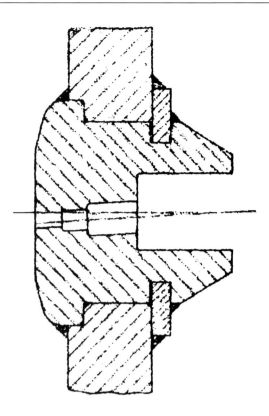

BELOW: Still retaining the double loop stowage for the track cable on the left superstructure side and **S-Mine** dischargers, this **Pz.Kpfw.Tiger** completed in August 1943 has factory-applied **Zimmerit** anti-magnetic coating. (MJ)

Redesigned **Sehschlitzplatten** (vision slit plates) for both turret sides with a smaller profile and a wider vision slit to increase the lateral field of view, were introduced in August 1943.

Chapter 3: Panzerkampfwagen Tiger Ausf.E

Several minor modifications are present on this **Pz.Kpfw.Tiger Ausf.E** completed in September 1943, including reversal of the shovel mounts on the glacis, welding a tab to the glacis for securing a chain to the machinegun ball mount plug, deletion of the right headlight, and reversal of the direction in which the last two **S-Mine** dischargers were fired. (BA)

Panzerkampfwagen Tiger Ausf.E – Fgst.Nr.250570 completed in October 1943 – dropped toggle bolts by machinegun ball mount and added tab on the glacis to secure a chain for a plug – single headlight mounted in center – dropped **S-Mine** – **Gleitschutzpickeln** (chevrons) on track li face – starter crank handle stowed by antenna base – new turret with simplified turret ring and cast cupola – pistol port plug – wider slit in lowe profile vision slits on turret sides – escape hatch lid sides no longer beveled – ventilation fan centered above gun breech – wider armor guard ov loader's periscope

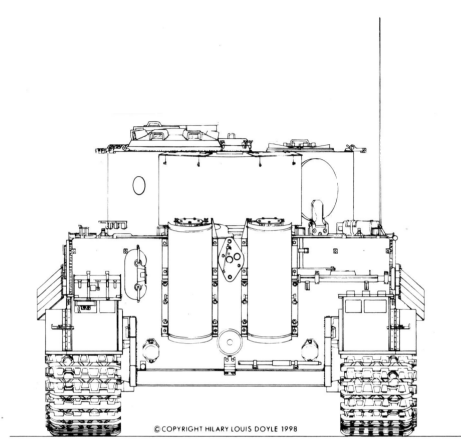

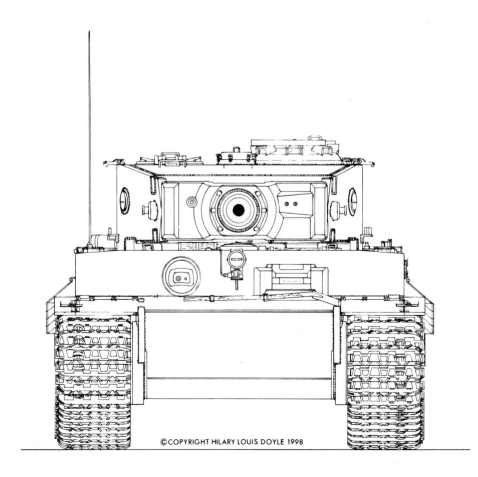

Chapter 3: Panzerkampfwagen Tiger Ausf.E

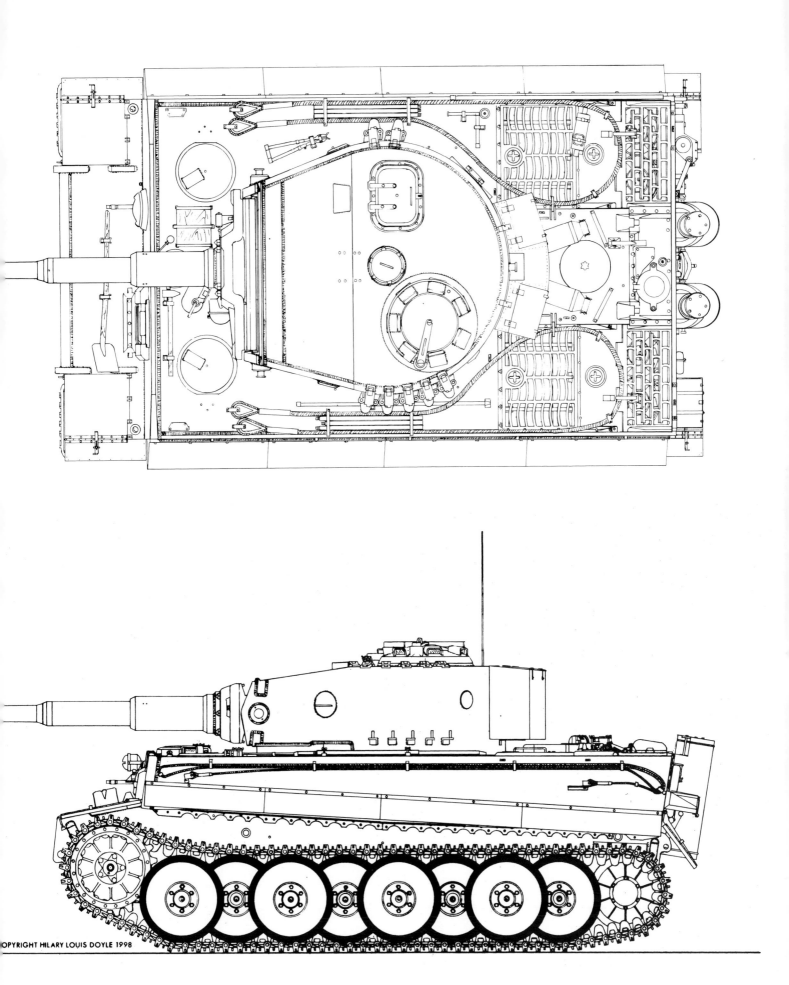

These two Tigers, completed in October 1943 and issued to **s.Pz.Abt.501**, still have fittings for **Feifel** air cleaners on the rear deck and pistol port plugs in the left turret side. (WS)

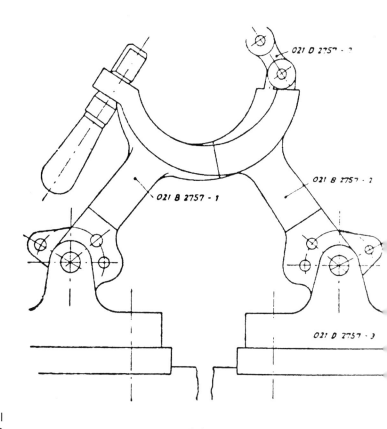

A **Heckzurrung** (rear travel lock) was mounted on the right rear hull (starting with **Fgst.Nr.250635** in November 1943 and stopping with **Fgst.Nr.250875** in February 1944).

Chapter 3: Panzerkampfwagen Tiger Ausf.E

Tiger I (**Fgst.Nr.250641** completed in November 1943) was one of the first with an external travel lock. The unit did not weld a bar across the front of the hull to carry spare track links. (WS)

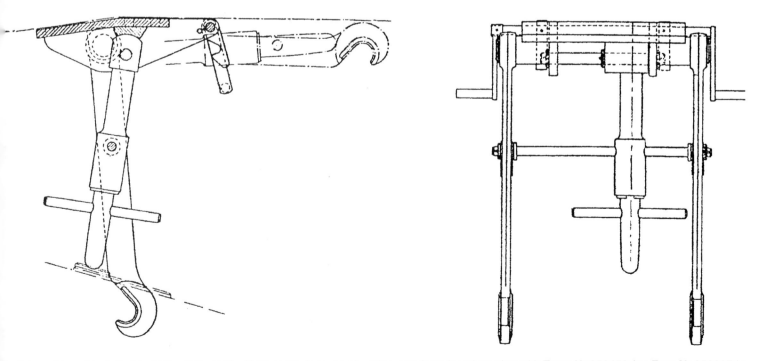

An internal travel lock that held the **8.8 cm Kw.K.36** at 15 degrees elevation was introduced starting with **Turm Nr.250450** (on **Fgst.Nr.250697** in December 1943).

Starting with **Fgst.Nr.250635**, Tigers (such as this one completed in December 1943) had a rear travel lock. Fittings to mount a C-hook on the hull rear had been added starting in September and the track tool stowage box had been dropped in late October 1943. The rectangular canvas rain guard has been erected over the commander's cupola. (BA)

Completed in December 1943, this Tiger issued to **s.SS-Pz.Abt.101** has a rear travel lock, still has fasteners for mounting a shovel across the glacis, and has the headlight mounted in the center of the driver's front plate (a modification which was initiated in October 1943). (BA)

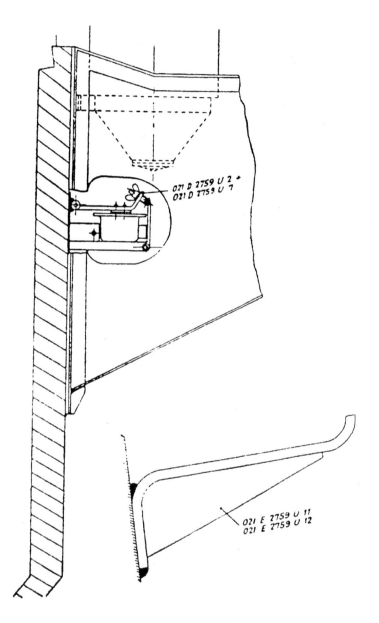

The higher lifting capacity **20 t Winde** (20 ton jack) was introduced and the mounting brackets relocated starting with **Fgst.Nr.250772** (produced in January 1944).

GERMANY'S TIGER TANKS - D.W. to TIGER I

Panzerbefehlswagen Tiger Ausf.E completed in January 1944 – dropped shovel on glacis, Feifel air cleaners on rear deck, and track tool stowage box on hull rear – mounted external travel lock on top right of hull rear – changed to 20 ton jack – flexible antenna base on turret roof for 2 meter rod – flexible antenna base on left rear deck for 1.4 meter rod – flexible antenna base for **Sternantenne** in large cylindrical guard for insulator – holder for **Sternantenne** extensions on right hull side – exhaust port in deck for **GG400** electrical generator set – tube on hull rear for stowing **Sternantenne**

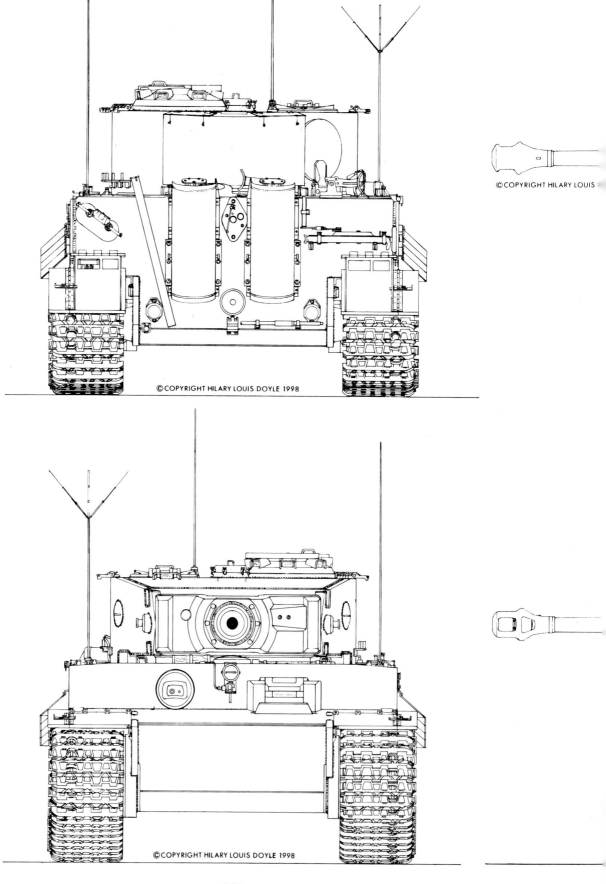

Chapter 3: Panzerkampfwagen Tiger Ausf.E

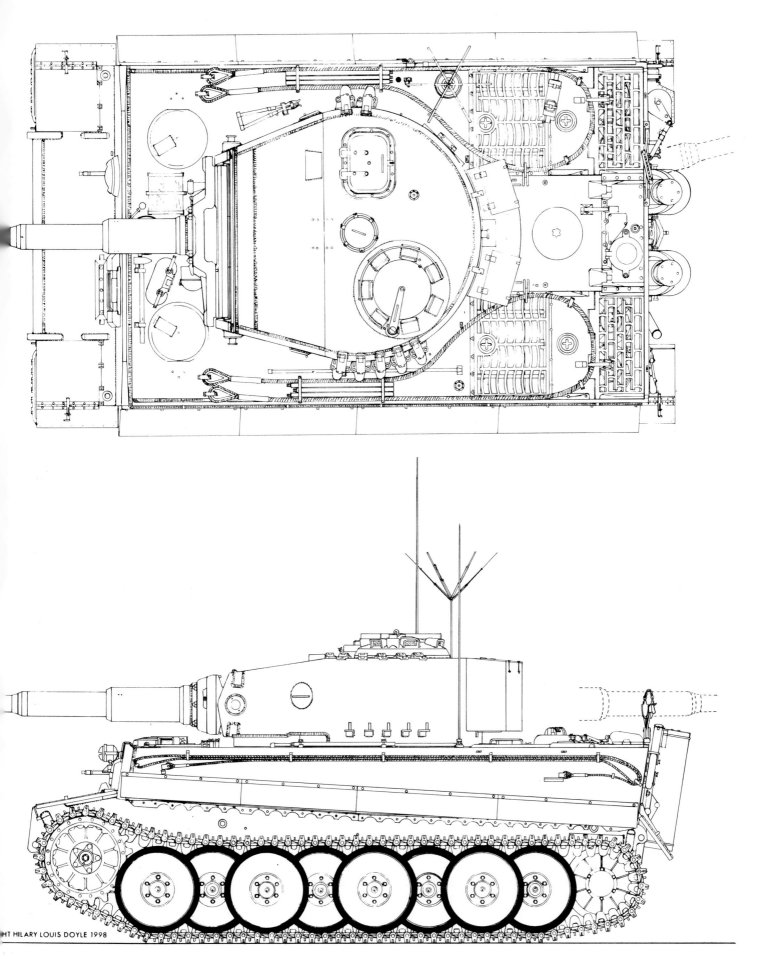

GERMANY'S TIGER TANKS - D.W. to TIGER I

THIS PAGE AND OPPOSITE: This **Pz.Kpfw.Tiger Ausf.E** no longer has the mountings for **Feifel** air cleaners on the hull rear and has a 20-ton ja (showing that it was completed after **Fgst.Nr.250772** in the latter half of January 1944). The face of the hull side extensions at the front have bee cut out to allow additional room for pivoting tow shackles with C-hooks. (BA)

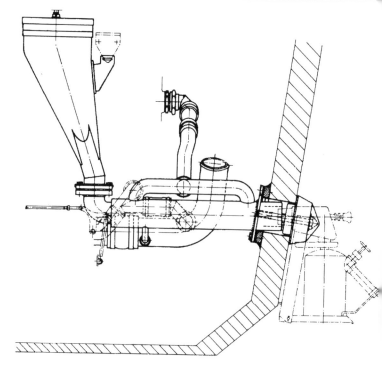

Starting with **Fgst.Nr. 250823** in February 1944, a hole was bored through the hull rear below the left muffler (externally covered with an armor cap held by two bolts) and a **Fuchsgeraet** (engine coolant heater) for cold starting installed. The engine coolant was heated using a blowtorch mounted on the outside.

142

Chapter 3: Panzerkampfwagen Tiger Ausf.E

THIS PAGE AND OPPOSITE: **Pz.Kpfw.Tiger Ausf.E** (**Fgst.Nr.250829** completed early in February 1944) is one of the first with steel-tired rubber saving roadwheels (which started with **Fgst.Nr.250822**). (BA)

Chapter 3: Panzerkampfwagen Tiger Ausf.E

145

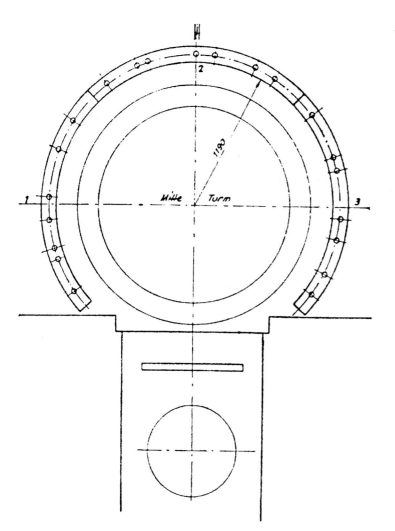

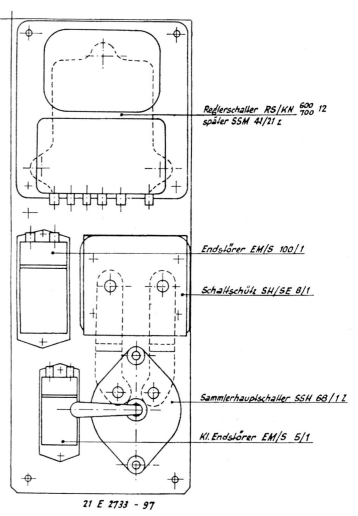

A **Turmfugenschutz** (turret ring guard) bolted to the superstructure deck at the base of the turret was introduced starting with **Fgst.Nr.250850** in February 1944.

Electrical components (including the voltage regulator, noise suppressors, starter protection fuzes, and master switch) were removed from inside of the hot engine compartment and mounted in the crew compartment on the vehicle-right side of the firewall, starting with **Fgst.Nr.25086** in February 1944.

OPPOSITE: This **Pz.Kpfw.Tiger Ausf.E** in Hungarian service was completed late in February 1944 after introduction of the turret ring guard (a **Fgst.Nr.250850**). While it doesn't have an external travel lock (dropped after **Fgst.Nr.250876**), it still has the base mounting plates welded on the top of the hull rear. (GF)

Chapter 3: Panzerkampfwagen Tiger Ausf.E

GERMANY'S TIGER TANKS - D.W. to TIGER I

These photographs, taken at the Henschel assembly plant, are of a **Pz.Kpfw.Tiger Ausf.E** completed in late February/early March 1944. It has the smaller 600 mm diameter idler, engine starter plate with mounting bolts, and tool stowage on the deck rearranged due to the turret ring guard. It still has a 25 mm thick turret roof with the deflector guard surrounding the loader's hatch. (KRP)

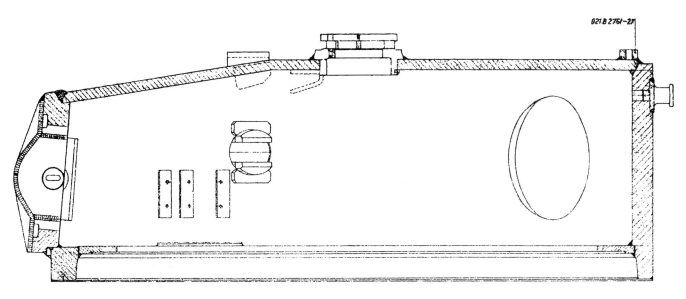

Starting with **Fgst.Nr.250991** in late March 1944, the turret roof armor was increased to 40 mm thick (with 15 mm protruding above the height of the turret sides).

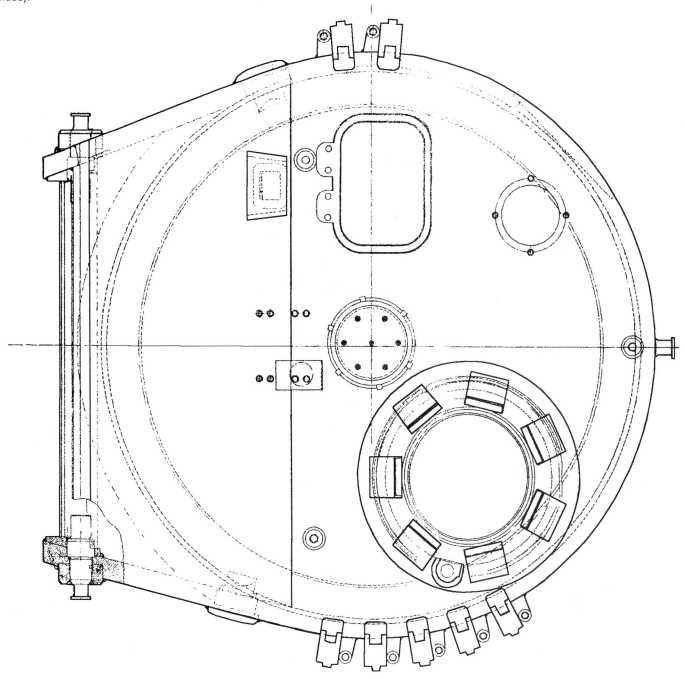

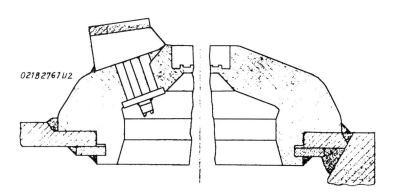

The armor casting for the commander's cupola remained the same as on the 25 mm thick roof (the 12-hour azimuth indicator ring had already been dropped in February 1944). The 40 mm roof was machined out by 15 mm and the base of the cupola casting welded to the turret roof.

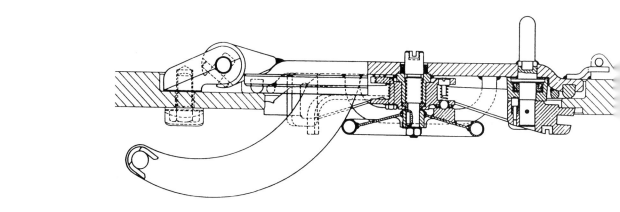

A new **Turmlukendeckel** (loader's hatch – part number 021 B 50606 was introduced with the 40 mm thick roof (in this case a forged lid from Krupp), countersunk into the turret roof with the handle moved off center to the right.

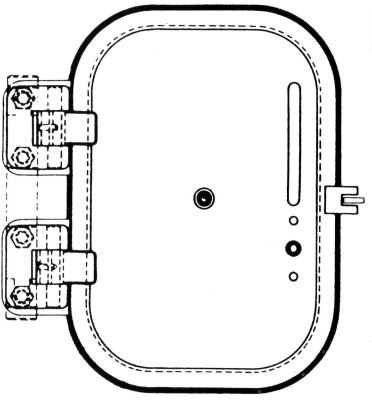

Chapter 3: Panzerkampfwagen Tiger Ausf.E

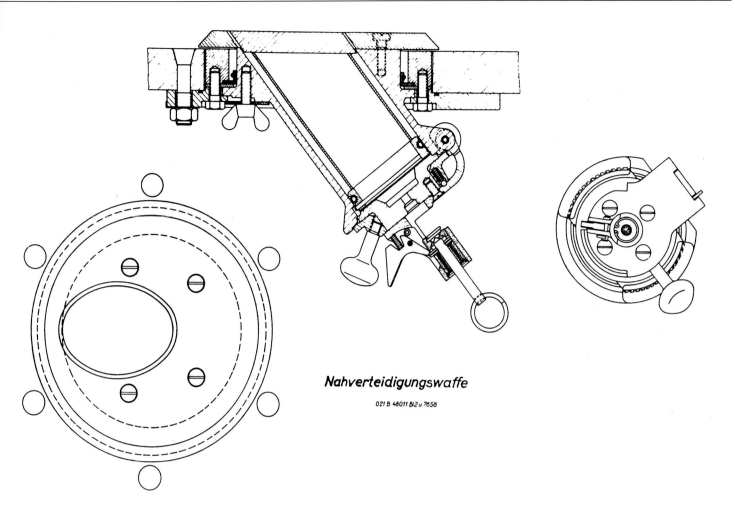

Nahverteidigungswaffe

021 B 48011 Bl.2 u.7658

The **Nahverteidigungswaffe** (close defense weapon) was mounted in the roof at the left rear (introduced at the same time as the 40 mm thick turret roof, starting in late March 1944 with **Fgst.Nr.250991**).

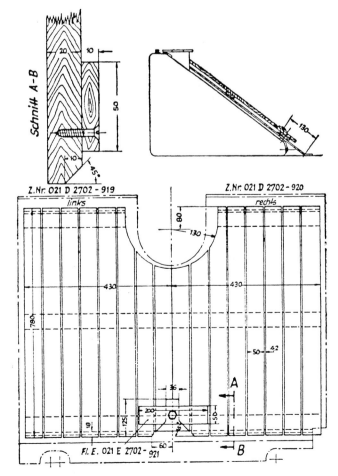

Wooden decking was installed over the top of the armor plate above the upper fuel tanks to prevent shell fragments and bullet splash from ricocheting and damaging the radiators (starting with **Fgst.Nr.251075** in April 1944).

Chapter 3: Panzerkampfwagen Tiger Ausf.E

OPPOSITE: This is a rare Tiger with 40 mm thick turret roof which still has a binocular **T.Z.F.9b** gunsight instead of a monocular **T.Z.F.9c** gunsight, which was also introduced in March 1944. (BA)

This Tiger I being used for training Hungarian crews has a 40 mm thick turret roof with a D.H.H.V. welded loader's hatch with long hinges behind which is mounted the **Nahverteidigungswaffe** (close defense weapon). (ECPA)

Exterior components on **Pz.Kpfw.Tiger Ausf.E** (**Fgst.Nr.251114** completed early in May 1944) (TLJ – Musée Blindes Saumur)

The machined drive sprocket wheel hub with sheet metal locking tabs to secure the bolts.

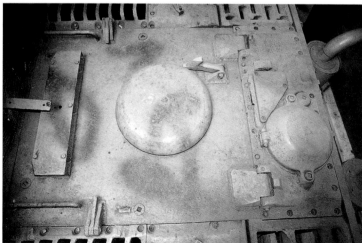

The 800 mm diameter rubber-cushioned steel-tired roadwheels common to the Tiger I, Tiger II, Panther II, and an experimental series of Panther Ausf.G.

The wooden decking over the fuel tank armor can be seen through the air intake louvers.

LEFT: The 600 mm diameter idler (minus the hubcap) with the track pin return plate repositioned due to the smaller diameter idler.

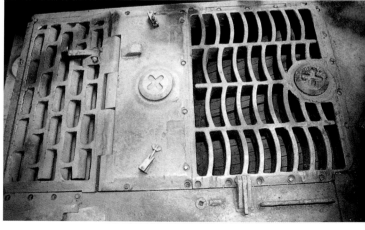

The casting had been standardized for the air exhaust louvers on both the right and left side (starting in April 1943).

LEFT: The hole has been plugged in the forged armor cap (from Krupp) which was formerly used to tighten a seal for submerged fording.

Chapter 3: Panzerkampfwagen Tiger Ausf.E

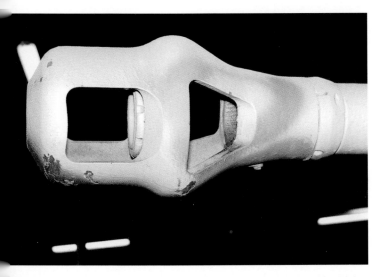

The standard muzzle brake outfitted to most of the **8.8 cm Kw.K.36 L/56** throughout its production run.

The tab on the glacis was for securing a plug (secured by a chain) for sealing the machinegun ball mount.

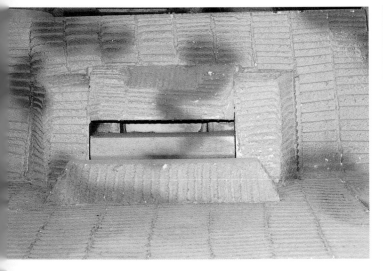

The driver's visor in the open position.

The single aperture in the gun mantlet for the monocular **T.Z.F.9c** telescopic gun sight.

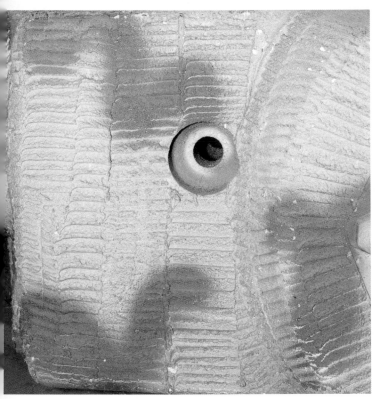

The emergency escape hatch with the hinge machined to clear the turret ring guard.

LEFT: The aperture in the gun mantlet for the coaxial machinegun.

155

Exterior components on **Pz.Kpfw.Tiger Ausf.E** (**Fgst.Nr.251114** completed early in May 1944) (TLJ – Musée Blindes Saumur)

The 40 mm thick turret roof was made from a single bent plate. One of the three "**Pilz**" (sockets for mounting a 2-ton jib boom) was located in front of the loader's hatch.

The base of the **Prismenspiegelkuppel** (commander's cupola with periscopes) 021St2762 was welded to the 40 mm roof.

The armor cap protecting the ventilation fan with a base ring with six tabs for securing the sheet metal cap used for submerged fording.

The 15 mm thick forged loader's hatch (from Krupp) was countersunk into the 40 mm thick turret roof.

Details of the clips used to hold the top of the spare track links onto the turret side.

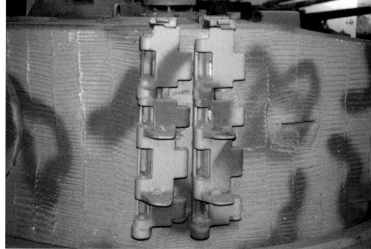

Two spare track links were stowed on the right turret side behind the loader's vision slit, which had been widened (starting in August 1943).

Chapter 3: Panzerkampfwagen Tiger Ausf.E

Interior of **Pz.Kpfw.Tiger Ausf.E** (**Fgst.Nr.251114** completed early in May 1944) (TLJ – Musée Blindes Saumur)

3-spoke wheel (introduced by August 1943) replaced the 5-spoke knurled wheel used to open and close the driver's visor. The **Argusenkapparat L.St.0.2** (steering device) with a 2-spoke steering arc is disconnected and upside down.

A **Zahnsegment-Turmzurrung** (toothed segment traverse travel lock with knurled knob) replaced the conical pin type (starting in July 1943). **Fgst.Nr.251114** is stamped into the side support between the ammunition racks on the right hull side.

The firewall, redesigned (starting with **Fgst.Nr.250501** in September 1943) with the electrical components clustered in the lower right corner (starting with **Fgst.Nr.250861** in February 1944)

The left side of the firewall with the cold start fuel injector

Ammunition racks in the left rear pannier behind the lower half of the gun counterbalance.

Ammunition racks in the left front pannier behind the gunner's seat (without cushion) and the elevation wheel with main gun trigger.

Interior of the turret on **Pz.Kpfw.Tiger Ausf.E** (**Fgst.Nr.251114** completed early in May 1944) (TLJ – Musée Blindes Saumur)

The modified coaxial machinegun mount, cushioned by a spring, was installed (starting with **Turm Nr.241** in May 1943). The reinforced **Bruecke** (bridge) 021B2761U3 with twice the number of bolts securing it to the turret ring (introduced starting in July 1943).

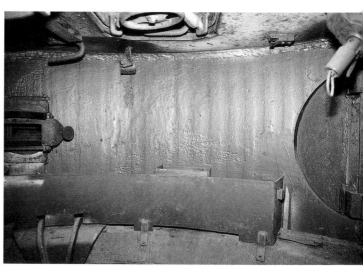

The right turret side from the vision block to the emergency escape hatch

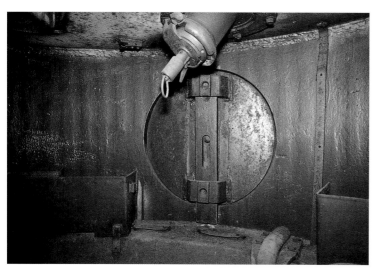

The breech-loaded **Nahverteidigungswaffe** (close defense weapon) (introduced starting in March 1944)

The rear of the turret with the top mount for the gun counterbalance spring.

The seven periscopes in the commander's cupola were held in place by tabs tightened with butterfly nuts. This is a **Prismenspiegelkuppel** (commander's cupola with periscopes) 021St2762 which did not have a 12-hour indicator ring.

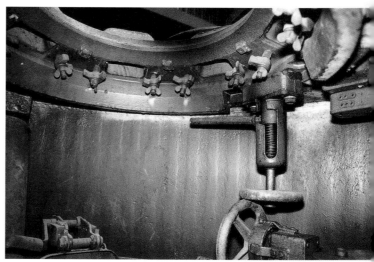

The wheel used to raise and lower the cupola hatch lid and the bar used to pivot the raised lid.

Chapter 3: Panzerkampfwagen Tiger Ausf.E

The auxiliary handwheel for the commander to help the gunner traverse the turret by hand. (This was not a commander's override as installed in American tanks.)

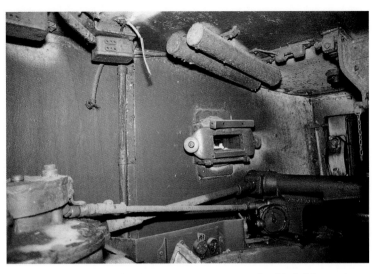

The 12-hour indicator for the gunner was mounted to his left front above the turret ring.

The left aperture in the gun mantlet has been plugged, while the rest of the mounting remained unchanged for the conversion to the monocular **T.Z.F.9c** telescopic gunsight.

The inside of the turret roof from the adjustable interior light to the loader's hatch.

The electric fan in the turret roof is missing behind the internal travel lock for the **8.8 cm Kw.K.36 R186**.

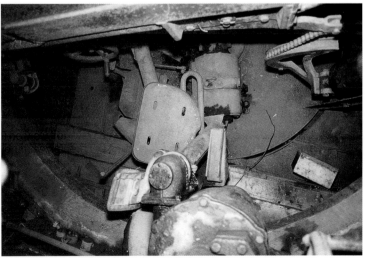

The view from the commander's cupola down to the gunner's seat (cushion missing) and the split foot pedal controls for power traverse.

159

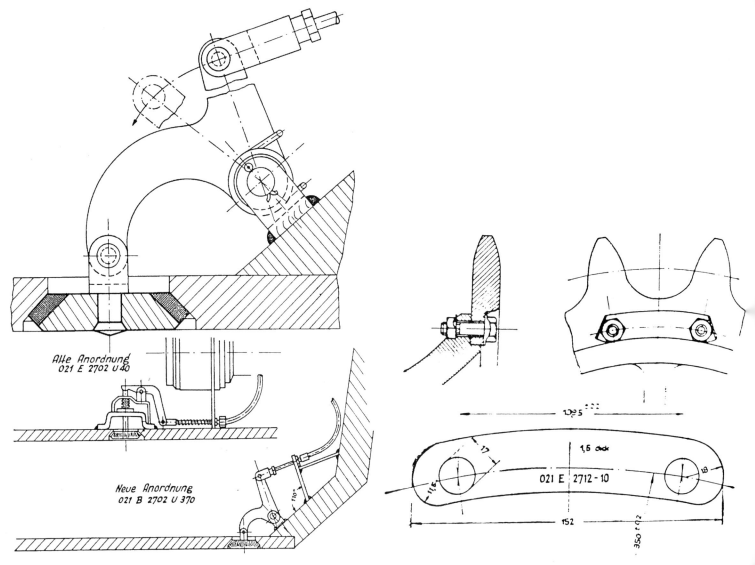

A simplified cable operated mechanism for remotely opening and closing the drain valve in the engine compartment was introduced starting with **Fgst.Nr.251165** in May 1944.

The ends of curved sheet metal stripes (held by two adjacent bolts holding the sprocket ring onto the **Kettentriebrad** -track drive wheel) were bent upward to keep bolt heads from rotating and loosening due to vibration, starting with **Fgst.Nr.251205** in early June 1944.

The turret roof was constructed using two 40-mm-thick plates and three slits were cut into the commander's cupola to drain off rainwater on this Tiger I (**Wanne Nr. 251113 amp** completed in May 1944) – now on display at Vimoutier, France. (HLD)

Chapter 3: Panzerkampfwagen Tiger Ausf.E

This Tiger I with the **s-Pz.Abt.510**, completed in late May/early June 1944, has three slits cut into the commander's cupola to drain off rainwater and still has the 3-bolt escape hatch hinge trimmed to clear the turret ring guard. (WS)

Panzerkampfwagen Tiger Ausf.E – Fgst.Nr.251227 completed in June 1944 – hull side extensions cut – steel-tired roadwheels – smaller 60 mm-diameter idler with wider/slanted track pin return plate – turret ring guard on deck – wooden decking over fuel tanks – mounting bolts on start alignment plate – port for **Fuchsgeraet** on lower left hull rear – **T.Z.F.9c** monocular gunsight – light weight muzzle brake with insert – two-piece 4 mm roof – countersunk loader's hatch – three **Pilze** sockets on turret roof – **Nahverteidigungswaffe** – drain slits in cupola – two bolts on shortene escape hatch hinge

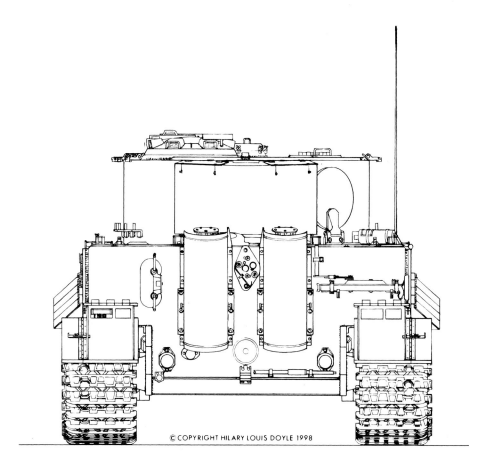

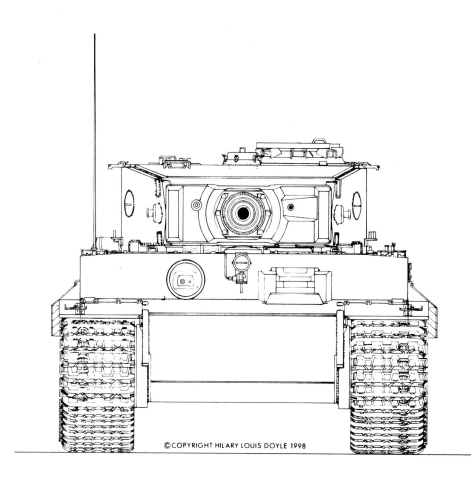

Chapter 3: Panzerkampfwagen Tiger Ausf.E

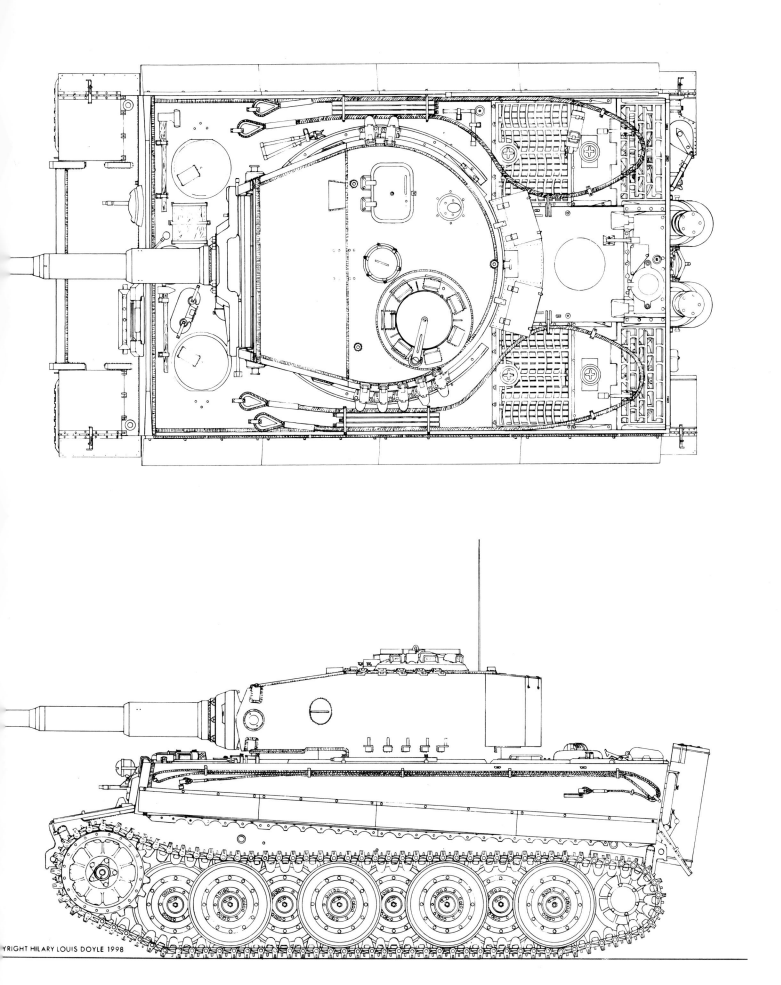

163

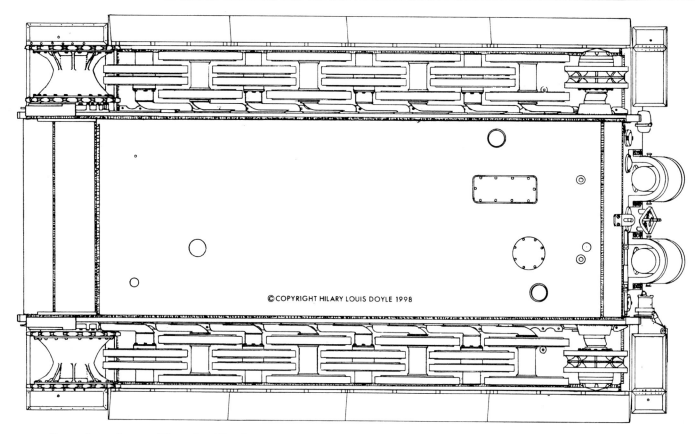

Panzerkampfwagen Tiger Ausf.E – Fgst.Nr.251227 completed in June 1944 – arrangement of steel-tired roadwheels

3.5 MODIFICATIONS AFTER ISSUE

The **Waffenamt** authorized the field units to implement a series of improvements on their Tiger I after issue. These included:

a. Starting in November 1942, add spring counterbalances to the **Ladeschuetzenluke** (loader's hatch) and **Kommandantenluke** (commander's hatch).
b. Starting in November 1942, weld support sleeves to the top of the cupola to support the **Regenschutzhaube** (rain shield) for the commander.
c. By February 1943, add another six bolts to tire retaining rings which have only six bolts.
d. Starting in February 1943, add a sheet metal guard to protect the commander from back flashes when spent casings are ejected.
e. Starting in May 1943, exchange 32 mm head bolts with 27 mm head bolts on the roadwheels to ensure that locking strips are long enough.
f. On 27 November 1943, the units were ordered to immediately reduce the maximum speed of the Maybach HL 230 engines. The engine was to be governed at about 2500 rpm under full load.
g. Starting in January 1944, apply **Zimmerit** anti-magnetic coating to Tigers that did not already have the coating applied at the assembly plant.
h. Starting in June 1944, weld three **Pilze** for **Befehlskran 2t** on the turret roof.
i. Starting in June 1944, cut off the bottom of the hinge for the turret escape hatch to ensure clearance after bolting a **Turmfugenschutz** (turret ring guard) to the hull.

j. Starting in July 1944, when steel-tired roadwheels are used to replace rubber-tired roadwheels, roadwheels directly in front of and behind that station are also to be replaced with steel tired roadwheels.
k. Starting in September 1944, paint two to three out of every ten **Verladekette** (track links) green in to order to distinguish Tiger I loading track from Tiger II loading track (to be painted red).
l. Starting in September 1944, extend the width of the track pin return plates on older Tigers by welding a wider section onto the bottom to within 270 mm of the center of the idler wheel axle.
m. Starting in October 1944, fabricate eight retaining strips out of sheet metal to be used for stowing an additional sixteen 8.8 cm rounds alongside the pannier racks above the lower ammunition bins.
n. Starting in October 1944, field units were ordered to stop applying **Zimmerit** anti-magnetic coating to Panzers.
o. Starting in November 1944, cut the connecting loop (over the sprocket tooth hole) out of the track links to improve self-cleaning and reduce track climbing.

Although it was officially discouraged, the troops themselves made several superficial modifications to the Tiger I. The most prevalent modification, performed by the units' maintenance section, was welding bars across the lower hull front to carry additional spare track links. In addition, many units carried spare track links across the driver's front plate. Toward the end of the war additional spare track links were mounted on the turret sides of some Tiger I in **s.Pz.Abt.507** and **510**.

Chapter 3: Panzerkampfwagen Tiger Ausf.E

This production series **Pz.Kpfw.VI (H) Ausf.H1** (**Fgst.Nr.250053**), completed in December 1942, was confiscated by **Wa Pruef 6** for testing and subsequently modified by cutting the hull side extensions for tow shackles and changing to the **gummigefederten Stahllaufrollen** (rubber-cushioned, steel-tired roadwheels). (WJS)

Pz.Kpfw.Tiger Ausf.E (**Fgst.Nr.250287**), completed early in June 1943, had the **Zimmerit** anti-magnetic coating applied by the troops using the exact pattern described in the modification order. (NA)

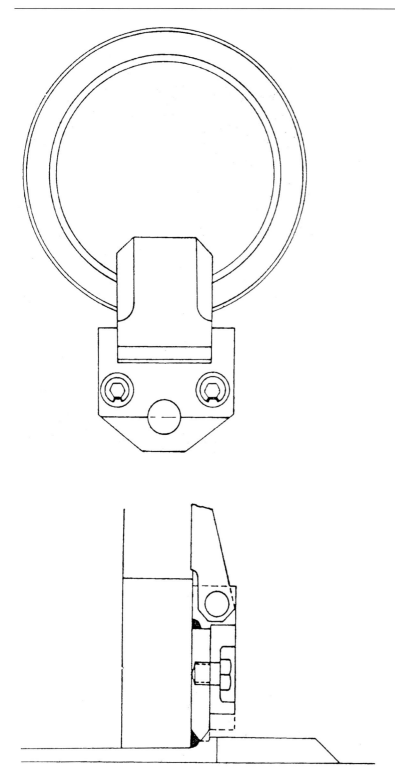

The bottom of the hinge holding the escape hatch had to be cut off to allow clearance when a **Turmfugenschutz** (turret ring guard) was bolted to the superstructure roof (modification printed in H.T.V.Bl. dated 1 July 1944).

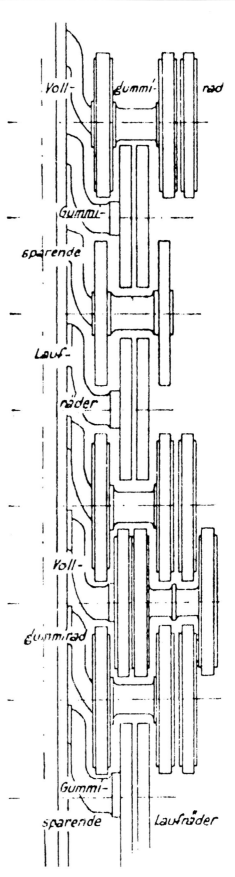

This sketch, printed in H.T.V.Bl. dated 15 July 1944, shows the correct method of mixing the newer **Gummisparende Laufraeder** (rubber-saving roadwheels) in with the older **Vollgummiraeder** (rubber-tired roadwheels).

Chapter 3: Panzerkampfwagen Tiger Ausf.E

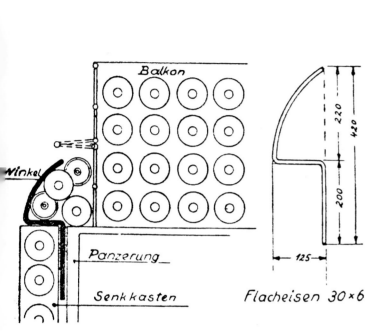

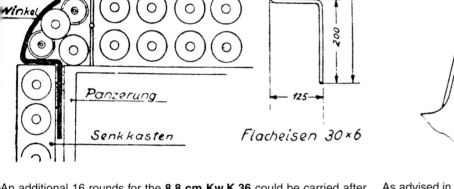

Kgs 63/725/130, Tiger E

An additional 16 rounds for the **8.8 cm Kw.K.36** could be carried after field units made and installed retaining clips as shown in H.T.V.Bl. dated 1 October 1944.

As advised in H.T.V.Bl. dated 15 November 1944, the **Buegel** (connecting loop) on each track link was to be cut off to improve self-cleaning and reduce track climbing.

Having been issued Tigers before tool stowage was standardized, only the **1.Kp./s.Pz.Abt.502** welded fasteners onto the base armor. Many modifications were made on the Tigers transported to Tunisia with the **1.Kp./s.Pz.Abt.501** including headlights moved to the front of the driver's front plate, a blade sight welded to the turret front, side extensions on rear track guards, muffler guards, spare track links on lower hull rear, and the tool box mounted on hull rear. Prior to February 1943, the **Pz.Kpfw.III** stowage bin on the rear of the turret was replaced with a larger stowage bin by most of the units (including **s.Pz.Abt.501**, part of **1.Kp./s.Pz.Abt.502**, **3.Kp./s.Pz.Abt.503**, **Grossdeutschland**, and all three **SS-schwere Kompanien**).

Lesser miscellaneous changes included sun/rain shields mounted above the gun sight apertures (on gun mantlets where this area hadn't been reinforced) by **s.Pz.Abt.503** in early 1943, barbed wire entanglements attached along the superstructure sides by **s.Pz.Abt.505** and later **s.Pz.Abt.501** in 1943, and logs carried on the superstructure side by **s.Pz.Abt.503** in 1943 and **s.Pz.Abt.505** in 1944. There were also isolated instances of individual Tigers being modified by welding a step and handle onto the front, mounting C-Hooks in brackets on the driver's front plate, welding brackets to carry a water can on the full rear, and removing the stop on the drum cupola so that the opened cupola lid lay flat.

Appendices

Appendices

APPENDIX A1 - Pz.Kpfw.VI (VK 4501 P) (Ausf.P)
Wa Pruef 6 dated 5 March 1942

Major Data
Combat Weight .... 52500 kg
Armor
Turret
Front .... 100 mm
Side .... 80 mm
Rear .... 80 mm
Deck .... 25 mm
Armament
    1 - 8,8 cm Kw.K. L/56
    2 MG 34
    1 MP
Traverse .... Hand and Hydraulic [sic]
Crew .... 5 Men

Further Data
Measurements
Gun Overhang .... 2.94 m
Overall Length .... 6.60 m
Overall Width
  (Loading Profile) .... 3.14 m
  (Cross Country) .... 3.24 m
Overall Height .... 2.92 m
Width Outside Tracks
  (Loading Profile) .... 3.14 m
  (Cross Country) .... 3.24 m
Track Width
  (Loading Profile) .... 0.50 m
  (Cross Country) .... 0.60 m
Wheel Base .... 2.64 m
Fording Ability .... Submerged
Ground Clearance .... 0.48 m
Step Climbing .... 800 mm
Observation .... Commander's Cupola
    Driver's Periscope
    T.Z.F.9b Gun Sight
    K.Z.F.2 MG Sight
Radio Sets .... Fu 5 and Fu 2
Direction Finding .... Kurskriesel (Course Compass)

APPENDIX A2 - PORSCHE "TYP 101A"
Porsche Document Sk.7949 dated 5 October 1942

Model	**101A**
Normal Crew Access	Turret
Emergency Escape	Turret Side
Main Gun	8,8 L/56
Ammunition	
Length mm	900
Weight Kg	19
Stored in Turret	0
Hull	50
on Floor	20
Number of Machineguns	1 MG34
Total Weight	59000
Track	
Pitch mm	130
Width mm	500/600/640
Length mm	4115
Ground Pressure	
Kg/cm^2	1.36/1.14/1.06
Suspension	
Model	101
Number of Springs	6
Roadwheels	12
Weight/Roadwheel	4650
Type of Drive Train	Electrical
	2 E Generator
	2 E Motor
Drive Sprocket Ø mm	794
Maximum Speed km/hr	35
Slope Climbing Ability	30°
Forward Idler	
Wheel Ø mm	794
Brakes	in wheel
Track Adjustment	Inside
Stopping Brakes	Hand lever
Auxiliary Motor	T 141
Engine Cooling	Air
Engine Model	101/1
Type	Gasoline
	Porsche
Number of Cylinders	10
Bore mm	115
Stroke mm	145
Volume liters	15
Speed rpm	2000 (2200)
Horsepower	270 (235)
Engines per Vehicle	2
Power Ratio HP/t	8

Appendices

APPENDIX A3 - PORSCHE DESIGN DATA
Dated 21 October 1942

Porsche Typ	101	102
Characteristic	Gasoline Electric	Diesel Hydraulic
Type	VK 4501 P	VK 4501 P
Overall Length		
Gun forward	9540 mm	=
Gun to the rear	7770 mm	=
Overall Width		
Combat Track	3320 mm	=
Loading Track	3140 mm	=
Overall Height	2900 mm	=
Length over Track	~6600 mm	=
Hull Type of Cutting	Mechanical	=
Length, outside/inside	6390/5610 mm	=
Width, outside/inside	2890/1800 mm	=
Height, outside/inside	1377/1225 mm	=
Width, Basket	1570 mm	=
Greatest Bore Distance	6130 mm	=
Ground Clearance	480 mm	=
Gauge	2680 mm	=
Obstacle Climbability	800 mm	=
Gun Firing Height	2200 mm	=
Engine (Type)	Petrol Air cooled	=
Torque at Maximum	2 x 99 mkg	=
Power Output	2 x 320 HP	=
Maximum Torque	2 x 108 mkg	=
	(n = 1800 rpm)	=
Fuel Consumption	270 gm/HP	=
Cooling Fan (Type)	Axial (2/engine)	=
Fan Power Consumption	2 x 40 HP	=
Air Volume	2 x 5.0 m³/sec	=
	50000 kg/h	=
Transmission (Type)	Petrol Electric	Petrol Hydraulic
Gear Change	Electric	Hydraulic
No. of Gears	Stepless	Two torque convertor steps
Total Reduction	about 1:110	about 1:132
No. of Main Gears	None	10
No. of Auxiliary Gears	None	4
Steering (Type)	Electric	BMM single radius hydrodynamic brakes
Intermediate Reduction	None	1:1.04
No. of Gears	None	13 main gears and 7 auxiliary for pumps
No. of Clutches	None	2 multi-plate
Final Drive (Type)	Epicyclic	=
No. of Gear Wheels	2 x 7	=
Total Gear Reduction in Bottom Gear	Generators in Series	Curves from Voith
Total Gear Reduction in Top Gear	Generators in Parallel	Curves from Voith
Total Number of Gear Wheels Engine to Track	2 x 7	40
Total Number of Clutches Engine to Track	2 slipping clutches	
Drive (Type)	Rear	=
Driving Sprocket	Cast steel with steel gear rim	=
Pitch Circle	794 mm	=
No. of Teeth	19	=
Height/Width	60/40	=
Track (Type)	Dry	=

$Torque at Maximum$ note: using mkg units as shown.

Track Support Length	4115 mm	=
Track Width	640 mm	=
Track Pitch	130 mm	=
Top Rollers (Type)	Twin Rollers	=
Number	6 per side	=
Outer Diameter	700	=
Width of Rubber	4 inner rubber rings, each 31 mm	
Rubber Rings D:d	633/535 mm	=
Suspension (Type)	Porsche Toggle	=
Arm Bearing	Taper roller	=
Spring (Type)	Torsion bar	=
Armor Thicknesses		
Turret Front	100 mm	=
Turret Side	80 mm	=
Turret Top	25 mm	=
Turret Rear	80 mm	=
Hull Front	100/80 mm	=
Hull Side	80 mm	=
Hull Top	25 mm	=
Hull Rear	80 mm	=
Hull Belly	20 mm	=
Total Weight	60 tons	=
Power/Weight Ratio	10.6 HP/ton	=
Center of Gravity	About 120 mm in front of center with gun forward	
Ground Pressure	1.13 kg/mm^2	=
Track Length/Gauge	1.53:1	=
Rubber Shear Stress	1.3 kg/cm^2	=
Max. Speed	About 35 km/h	=
Min. Speed	About 1.3 km/h	=
Max. Angle of Climb	35°	=
Fuel Tank Capacity	520 liters	
Bogging Ability	1.3	=
Turning Radius	On the spot	=
Radius of Action		
at 45 kg/t Resistance	about 105 km	
Radius of Action		
at 90 kg/t Resistance	about 48 km	
Radius of Action		
at 350 kg/t Resistance	about 15 km	
Ground pressure		
at 50 mm	1.03 kg/cm^2	=
at 100 mm	0.98 kg/cm^2	=
at 200 mm	0.89 kg/cm^2	=
Volume of Hull (Empty)	about 14.8 m^3	=
Volume of Fighting Compartment (Empty)	about 7.8 m^3	=
Volume of Fighting Compartment with Components	about 4.3 m^3	=
No. of Torsion Bars	6	6

Appendices

APPENDIX B1 - Pz.Kpfw.VI (VK 4501H) (Ausf.H1) (Tiger)
Wa Pruef 6 dated 5 March 1942

Major Data
Combat Weight .... 52000 kg
Speed
Road .... 40 km/hr
Average terrain .... 18-20 km/hr
Range
Road .... 140 km
Cross-country .... 85 km
Armor
Turret
Front .... 100 mm
Side .... 80 mm
Rear .... 80 mm
Deck .... 25 mm
Hull
Front .... 100 mm
Side .... 80 mm
Rear .... 100 mm (sic)
Deck .... 25 mm
Belly .... 25 mm
Armament
                                         1 - 8,8 cm Kw.K. L/56
starting with 101st .... 1 - 7,5 cm Kw.K. L/70
                                         2 MG 34
                                         1 MP
Traverse .... Hand and Hydraulic
Crew .... 5 Men

Further Data
Measurements
Gun Overhang .... 1.73 m
Overall Length .... 6.30 m
Overall Width
  (Loading Profile) .... 3.14 m
  (Cross Country) .... 3.55 m
Overall Height .... 2.90 m
Width Outside Tracks
  (Loading Profile) .... 3.14 m
  (Cross Country) .... 3.55 m
Track Width
  (Loading Profile) .... 0.520 m
(Cross Country) .... 0.725 m
Wheel Base .... 2.82 m
Fording Ability .... Submerged 4.0 m
Ground Clearance .... 0.47 m
Further Capabilities
Motor type .... Gasoline
Power .... 600 metric HP
Power/Weight Ratio .... 11.5 HP/ton
Step climbing ability .... 0.850 m
Slope climbing ability .... 30°
Fuel capacity .... 570 liters
Fuel consumption (estimated)
Road .... 400 l/100 km
Cross-country .... 700 l/100 km
Ammunition .... 96 rounds for main gun
                                         3000 rounds machinegun
                                         320 rounds for machinepistol

6.	Observation	Commander's Cupola
		Driver's Periscope (K.F.F.2)
		T.Z.F.9b Gun Sight
		K.Z.F.2 MG Sight
7.	Radio Sets	Fu 5 and Fu 2
		Intercom
8.	Direction Finding	Kurskriesel (Course Compass)

Appendices

APPENDIX B2 - PANZERKAMPFWAGEN TIGER I DATA
Wa Pruef 6 dated 14 December 1943
(Current as of 1 November 1943)

<u>Weights</u>
Combat Weight .... 54,000 kg
Turret Weight .... 11,000 kg

<u>Speeds</u>
Highest Speed .... 45 km/hr
Sustained Speed on Roads (Autobahn) .... 30 km/hr
Sustained Speed on Average Terrain .... 20-25 km/hr

<u>Range</u>
Road .... 120 km
Average Terrain .... 85 km

<u>Fuel Consumption per 100 km</u>
Road .... 450 l
Average Terrain .... 650 l

<u>Fuel Capacity</u> .... 540 l

<u>Measurements</u>
Length Overall
    with Gun Forward .... 8.451 m
    with Gun Aft .... 8.379 m
    without Gun Overhang .... 6.280 m
Gun Overhang with Gun Forward .... 2.171 m
    Width Overall with Cross-Country Tracks .... 3.560 m
    Wheel Base .... 2.822 m
    Track Width .... 0.725 m
Width Overall with Rail Loading Tracks .... 3.230 m
    Wheel Base .... 2.622 m
    Track Width .... 0.620 m
Free Width Inside Hull .... 1.800 m
Free Diameter Inside Turret Ring .... 1.820 m
Height Overall .... 3.000 m
Turret Height with Commander's Cupola .... 1.200 m
Height of Centerline of Gun at 0° Elevation .... 2.195 m
Ground Clearance .... 0.470 m
Track Contact Length/Wheel Base 1.28
Ground Pressure .... 1.03 kg/cm^2
    with Tracks Sinking in 20 cm .... 0.90 kg/cm^2

<u>Capabilities</u>
Engine Type .... Gasoline
Engine Make .... HL 230 P45
Power .... 700 HP
Power to Weight Ratio .... 13 HP/ton
Trench Crossing Ability .... 2.3 m
Step Climbing Ability .... 0.79 m
Grade Climbing Ability .... 35°
Fording Ability .... 1.70 m
Submerged Fording Ability .... 4.0 m

<u>Armor</u>
Turret
    Gun Mantlet .... 120 @ 0°
    Front .... 100 @ 0° (Sic)
    Side .... 80 @ 0°
    Rear .... 80 @ 0°
    Roof 25
Hull and Superstructure
    Driver's Front .... 100 @ 9°
    Lower Hull Front .... 100 @ 15° (Sic)
    Side .... 80 @ 0°
    Rear .... 80 @ 9°

	Roof	25
	Belly	25

9. <u>Armament</u>
   1 - 8.8 cm Kw.K. L/56
   1 - M.G. in turret, 1 M.G. in hull
   1 - M.P.
   1 - Smoke Discharger Device
10. <u>Ammunition</u>
    A.P. and H.E. 92
    Machinegun 4500
    Machinepistol 192
11. <u>Crew</u>
    5 men; 1 commander, 1 gunner, 1 loader, 1 driver, and 1 radio operator
12. <u>Observation and Aiming Devices</u>
    Cupola for commander
    T.Z.F.9b gun sight for gunner
13. <u>Traverse and Elevation</u>
    +18 to -9° arc of elevation, 360° traverse
14. <u>Direction Finding Device</u>
    Kurskriesel (Course Compass)
15. <u>Radio Sets</u>
    Ultra-Shortwave Fu.2 receiver
    Ultra-Shortwave Fu.5 receiver and transmitter
16. <u>Other</u>
    Bordsprechanlage 1 (intercom)

Appendices

APPENDIX B3-TIGER E TECHNICAL SPECIFICATIONS
by Henschel dated 9 February 1944

General Vehicle Data

Combat Weight .... 57,000 kg
Rail Loading Weight .... 52,500 kg
Speeds
Sustained Speed
    on Roads (Autobahn) .... 40 km/hr
    on Average Terrain .... 20-25 km/hr
Range on Roads (Autobahn) .... 195 km
    on Average Terrain .... 110 km
Capabilities
Trench Crossing Ability .... 2.50 m
Step Climbing Ability .... 0.79 m
Slope Climbing Ability .... 35°
Fordability .... 1.60 m
Submersion .... 4 m
Crew .... 5 men

Vehicle Measurements

Length Overall
    with Gun Forward .... 8,450 mm
    with Gun Aft .... 8,434 mm
    without Gun Overhang .... 6,316 mm
Gun Overhang with Gun Forward .... 2,116 mm
Width Overall, over Track Guards .... 3,705 mm
Height Overall .... 3,000 mm
Length over the Tracks .... 5,850 mm
Width over the Tracks
    with the Cross Country Tracks .... 3,547 mm
    with the Rail Loading Tracks .... 3,142 mm
Ground Contact Length of the Tracks .... 3,605 mm
    by Sinking in 20 cm .... 5,150 mm
Width between Track Centers (Wheelbase)
    with the Combat Tracks .... 2,822 mm
    with the Rail Loading Tracks .... 2,622 mm
Height from Ground to Top of Chassis .... 1,800 mm
Firing Height of the Main Gun .... 2,195 mm
Ground Clearance
    Front .... 470 mm
    Rear .... 470 mm

Hull with Superstructure

Largest Outside Length of the Hull .... 5,965 mm
Hull Width Outside
    at Suspension Height .... 1,920 mm
    at Superstructure Middle .... 3,140 mm
Hull Width Inside
    at Suspension Height .... 1,800 mm
    at Superstructure Middle .... 2,980 mm
Height of the Hull from Belly to the Deck .... 1,335 mm

Hull Armor .... <u>Thickness at Angle</u>
Driver's Front Plate .... 100 mm at 9°
Hull Front Lower .... 100 mm at 25°
Hull Side Lower .... 60 mm at 0°
Hull Rear .... 80 mm at 9°
Deck .... 25 mm at 90°
Belly .... 25 mm at 90°

Weight of Hull with Deck Ready for Assembly .... 20,800 kg

Suspension

	Cross-Country Track	Loading Track
Number of Guides per Track Link	2	2
Track Pin Length	716 mm	658.5 mm
Track Pin Diameter	28 mm	28 mm
Ground Pressure without Sinking In	1.05 kg/cm^2	1.46 kg/cm^2
Sinking In 20 cm	0.735 kg/cm^2	1.02 kg/cm^2
Steering Ratio (Track Contact Length divided by Wheel Base)	1.278	1.384

Roadwheels

Type of Suspension	**Geschachtelt** (interleaved)
Type of Roadwheels	Steel tires with rubber cushioning
Number of Roadwheels per Side	8
Roadwheel Diameter	800 mm
Weight Carried per Roadwheel	3,440 kg
Axle Distance from Roadwheel to Roadwheel	515 mm
Play between Roadwheel and Track Guide	2 mm

Suspension Arms and Suspension

Material for Bushings	Novetext
Type of Suspension	Torsion Bars
Number of Torsion Bars	16
Torsion Bar Diameter	58 mm front and rear, 55 mm middle
Torsion Bar Head Diameter	80 or 85 mm
Active Torsion Bar Length	1,730 mm
Total Torsion Bar Length	1,890 mm
Distance from Hull Bottom to Center of Bar	95 mm

Drive Wheel, Idler Wheel, and Shock Absorber

Effective Diameter of the Drive Wheel	841.37 mm
Segment Spacing for the Drive Wheel	131 mm
Diameter of the Idler Wheel	600 mm
Largest Distance for Track Adjustment	115 mm
Number of Shock Absorbers	4

Automotive System

Engine Designer	Maybach-Motorenbau
Type	HL 230 P45
Power at 3000 rpm, 20°C air temperature, and 760 mm air pressure	700 HP
Engine Weight	1,300 kg
Number of Cylinders	12
Stroke	145 mm
Bore	130 mm
Swept Volume	23,000 cm^3
Compression Ratio	1 to 6.8
Ignition	2 Magnetos with built-in **Zuendfunkenschnapper**
Fuel Consumption per 100 kilometers	
on Roads	270 l
on Average Terrain	480 l

Engine Cooling System

Type of Cooling	Water Cooled
Number of Radiators	2
Radiator Block Height	490 mm
Radiator Block Width	892 mm
Radiator Block Depth	200 mm
Area of Radiator Face	0.437 m^2
Number of Fans	2

Appendices

Diameter of the Fan Wheel	437 mm
Speed of the Fans at Maximum Engine Speed	
in the Summer	4,150 rpm
in the Winter	2,950 rpm
Type of Fan Drive	Universal Gears with Connecting Shaft
Maximum Power Needed for Fans	50 HP
Number of Air Filters	2
Manufacturers of Air Filters	Mann & Hummel

Transmission and Steering Gears

Length of Connecting Drive Shafts	
Front	885 mm
Rear	840 mm
Transmission Type	Maybach OG 40 12 16 A
Manufacturers	Maybach and Zahnradfabrik
Number of Forward Gears	8
Number of Reverse Gears	4
Overall Length of the Transmission	1,372 mm
Overall Width of the Transmission	556 mm
Overall Height of the Transmission	591 mm
Vehicle Speed with Engine Speed at 3000 rpm	
1. Gear	2.84 km/hr
2. Gear	4.34 km/hr
3. Gear	6.18 km/hr
4. Gear	9.17 km/hr
5. Gear	14.1 km/hr
6. Gear	20.9 km/hr
7. Gear	30.5 km/hr
8. Gear	45.4 km/hr
Reverse	3.75 km/hr
Overall Gear Reduction of the Transmission	1 to 16
Gear Cutting in the Universal Gear	Spiral
Gear Reduction in the Universal Gear	1 to 1.06
Steering Gear Manufacturer	Henschel & Sohn
Steering Gear Type	Double Radius L600 C
Gear Reduction of the Steering Gear	1 to 1.333
Number of Steering Stages	2
Number of Steering Clutches	4
Material for Clutch Facing	Jurid or Emero
Number of Gears	29
Minimum Turning Radius	3.44 m
Largest Turning Radius	165 m
Steering Controls	Argus Steering Device
Total Length of Transmission	
with Steering Gear	1,812 mm
Total Weight of Transmission with Steering Gear	1,345 kg

Final Drive and Brakes

Gear Reduction of the Final Drive	1 to 10.7
Type of Brakes	Argus Disk
Outer Diameter of the Brake Drum	550 mm
Material for the Brake Lining	Emero lining
Brake Cooling	Cooling Ribs on Housing
Type of Brake Operation	Mechanical Foot and Hand Levers

Volumes

Fuel Capacity	540 l

Traversing Turret

Turret Weight	11,000 kg
Turret Height with Commander's Cupola	1,200 mm
Free Inner Turret Ring Diameter	1,830 mm

Armament

8.8 cm Kw.K.36 (L/56)
Ammunition 92
M.G.42 (in the interim M.G.34)
39 Sacks with 150 Rounds of Belted Ammunition 4,800
M.P. 1
6 Magazines, each with 32 Rounds 192
Explosive Charges 3

When the new turret (with the pivoting hatch lid on the commander's cupola) is mounted, the following data apply:
Length Overall ....
    with Gun Forward .... 8,455 mm
    with Gun Aft .... 8,411 mm
    without Gun Overhang .... 6,335 mm
Gun Overhang with Gun Forward .... 2,122 mm
Overall Height .... 2,885 mm

Appendices

APPENDIX C - PART NUMBERS

As shown in the following list (extracted from a **Gruppenliste** dated 21Oct42), the turret mounted on the **VK 45.01 (H)** with **Gruppen-Nummer Serie 2760 - 2799** was originally designed for the **VK 45.01 (P)** with **Gruppen-Nummer Serie 860 - 899**.

	VK 45.01 (P)	VK 45.01 (H)
Turm Gruppe (Turret Group)	860 b	2760
Turmgehaeuse (Armor Body)	869 b	2760
Turmkugellager (Ball Bearing Race)	861	861
Kommandantenkuppel (Commander's Cupola)	862	862
Walzenblende (Gun Mantle)	863	863
Walzenabdichtung (Seal for Gun Mantle)	864	864
Lukendeckel (Loader's Hatch)	865	865
Turmschwenkwerk (Traverse Gear)	866	2766
Hoehenrichtmaschine (Elevation Gear)	867	867
MG Lagerung (Machinegun Mount)	868	868
Dichtstopfen f. M.G. (Sealing Plug for MG Aperture)	869	869
M.G.-Abzug (MG Firing Mechanism)	870	2770
Gurtsackhalterrahmen (MG Ammo Belt Sack Frames)	37462	37462
Lagerung f. Optik (Gunsight Mount)	2771	2771
Turmzurrung (Traverse Lock)	872	872 (Turm Nr.1-36)
12-Uhr-Zeigerantrieb (Azimuth Indicator Drive)	2773	2773
Richtungszeiger (Direction Indicator)	874	2774
MP-Klappe (Pistol Port)	875	875
Huelsenrutsche (Cartridge Slide)	877	N/A
Steuerung zum Fluessigkeitsgetriebe (Control for Hydraulic Drive)	N/A	2777
Dichtstopfen f. Huelsenrutsche (Sealing Plug for Cartridge Slide)	878	N/A
Fluessigkeitsgetriebe (Hydraulic Drive)	N/A	2778
Federausgleicher (Spring Counterbalance)	879	879
Turmzurrung (Traverse Lock)	880	880 (Ab Turm Nr.37)
Kommandantensitz (Commander's Seat)	5973	884
Schuetzensitz (Gunner's Seat)	885	885
Ladesitz (Loader's Seat)	886	886

Frischluftzufuehrung (Fresh Air Intake)	887	N/A
Wiegenabdichtung (Carriage Seal)	888	888
Sehschlitzplatte (Vision Slit Plate)	889	889
Zubehoerlagerung (Equipment Stowage)	890 b	2790
Schwenkregler (Traverse Control)	893	
Schleifringuebertrager (Swivel Contacts)	894	
Verteilerkasten (Junction Box)	896	
Elektr. Einrichtung (Electrical Equipment)	897	2793
Entlufterhaube (Exhaust Fan Cover)		2760/375
Notaussteig (Emergency Escape Hatch)		2760/452
Kommandantensitz (Commander's Seat)		2763
12-Uhr Zeigerantrieb (Aximuth Indicator Drive)		2764
Drehbuehne (Turret Platform)		2779

Drawing/part numbers with a "b" suffix apply to the new higher sided turret. The original eight turrets for the **VK 45.01 (P)** with lower sides had the same drawing/part numbers without the "b" suffix.

Only a partial list of the part numbers for the **VK 45.01 (H)** chassis (**Gruppen-Nummer Serie 2700-2759**) could be reconstructed from original documents (a spare parts manual has not been found in any archives). As can be seen from the following list, many components were taken over from its predecessor the **VK 36.01** (**Gruppen-Nummer Serie 4200-4299**).

Wanne (Armor Hull)	2701
Laufrad (Roadwheel)	4205
Leitrad (Idler Wheel)	4208
Seitenvorgelege (Final Drive)	4211
Verladeketten Kgs 63/520/130 (Loading Tracks)	4213
Lenkgetriebe Henschel L600C (Steering Gear)	4217
Luefter (Fan)	2724
Kraftstoffbehaelter (Fuel Tank)	2726
Lenkrad (Steering Wheel)	4228
Fahrerlukendeckel (Driver's Hatch)	2730
Turmantrieb (Turret Traverse Drive)	2742
Kuehlanlage (Cooling System)	2756
Fahrersehklappe (Driver's Visor)	9571
Kugelblende 100-9° (MG Ball Mount)	2782

Appendices

APPENDIX D - ARMOR SPECIFICATIONS

ARMOR FROM KRUPP

The 50 to 100 mm thick armor plates for the **VK 30.01**, **VK 36.01** and **VK 45.01(P)** were fabricated by Krupp in accordance with their specifications for **PP793** rolled armor plate. The tensile strength specified for 55 to 80 mm thick plates was 100 to 115 kg/mm^2 (equivalent to Brinell Hardness Number range of 294 to 338) and for 85 to 120 mm thick plates was 90 to 105 kg/mm^2 (equivalent to Brinell Hardness Number of 265 to 309). Krupp made the 25 mm thick roof plates out of **PPM942 (E112)** rolled armor plate, and the gun mantlet was made as a casting using alloy **FKM45**.

A decision had been made by February 1942 for Krupp to harden the face of the armor to increase resistance to penetration introduced in the **VK 45.01 (P)** starting with turret number 150019 and hull number 150050). The specification for the hardened face was set at 200 kg/mm^2 (equivalent to Brinell Hardness Number of 555) to a depth of 4 to 6 mm. The hardness specification for the interior and back of face-hardened plates remained unchanged from that specified for homogeneous armor plate.

In May 1942 a decision was made to replace **PP793** with a richer alloy, named **PP7182**. **PP7182** was to have the same composition of carbon, manganese, silicon, and molybdenum as **PP793**, but with slightly less chromium (1.7 to 2.3%) and enriched with scarce nickel (0.40 to 1.00%). The combined percentage of chromium and nickel in **PP7182** was to be at least 2.7%.

Specifications for the Krupp armor alloys were:

Armor Type	Spec. Date	Thickness	Resistance Kg/mm^2	Brinell Hardness	Alloy Percent
PPM942	Dec41	16-30 mm	115-130	338-382	1.75-2.55
PP793	1939	30-50 mm	110-125	324-368	2.25-3.20
PP793	1939	55-80 mm	100-115	294-338	2.25-3.20
PP793	1939	85-120 mm	90-105	265-309	2.25-3.20
PP7182	May42	85-120 mm	90-105	265-309	2.85-4.30
FKM45		All	85-100	250-294	2.35-3.60

Composition of each of the Krupp armor alloys is shown in the following table:

Armor Steel Alloy Composition in Percentage

	C	Mn	Si	Cr	Ni	Mo	V	P&S
PP793	0.32-0.42	0.30-0.60	0.15-0.50	2.00-2.40	—	0.15-0.30	—	<0.05
PPM942	0.44-0.54	0.80-1.10	0.50-0.80	0.80-1.10	—	0.15-0.35	<0.15	<0.05
PP7182	0.32-0.42	0.30-0.65	0.15-0.50	1.70-2.30	0.40-1.00	0.20-0.35	<0.15	<0.05
FKM45 Cast	0.35-0.45	0.60-1.10	0.20-0.60	1.50-2.00	—	0.15-0.25	0.10-0.25	<0.05

ARMOR FOR THE VK 45.01 (H) FROM D.H.H.V.

Dortmund Hoerder Huttenverein (D.H.H.V.) used their own alloys for rolled armor plate. When D.H.H.V. started manufacturing **VK 45.01(H)** armor hulls, alloy **C74** was used for plates up to 50 mm thick and alloy **C81** for plates over 50 mm thick. After August 1942, D.H.H.V. made the 100 mm thick frontal armor plates out of alloy **HB75**. Records have not been found revealing the original alloy used by D.H.H.V. for casting the gun mantlet, drive wheel, rear deck louvers, etc.

Specifications for the D.H.H.V. armor alloys were:

Armor Type	Spec. Date	Thickness	Resistance Kg/mm^2	Brinell Hardness	Alloy Percent
C74	1941	16-30 mm	115-130	338-382	2.35-3.25
C81	1941	55-80 mm	100-115	294-338	3.40-4.30
C81	1941	85-120 mm	90-105	265-309	3.40-4.30
HB75	Aug42	85-120 mm	90-105	265-309	2.50-3.35

Composition of each of the D.H.H.V. armor alloys is shown in the following table:

One of the two armor hulls produced by D.H.H.V. for the express purpose of test firing to determine the design's ability to withstand hits. The **Vorpanzer** (spaced armor), intended to defeat both tungsten-carbide and shape-charged armor-piercing projectiles, is shown in its lowered position. (TTM)

Appendices

	_C_	Mn	Si	Cr	Ni	Mo	V	P&S
274	0.44-0.51	0.55-0.85	0.20-0.50	1.40-1.80	—	0.40-0.60	—	<0.05
281	0.41-0.49	0.55-0.85	0.20-0.50	2.30-2.70	—	0.55-0.75	—	<0.05
HB75	0.32-0.42	0.30-0.65	0.15-0.50	2.00-2.40	—	0.20-0.30	—	<0.05

Armor Steel Alloy Composition in Percentage

STANDARDIZED ARMOR FROM ALL SUPPLIERS

During the production run of the Tiger I, Wa Pruef 6 gradually replaced the alloys invented by the individual armor suppliers alloys with standardized (**Einheits**) alloys that were to be produced by all armor suppliers. Thinner 25 mm plates were made from **E32** armor steel, the 60 mm to 80 mm plates from **E22** armor steel, and cast armor components from alloy "**A**" or later "**B**". However, 100 mm plates were still made out of **PP7182** by Krupp and **HB75** by D.H.H.V. to the end of the Tiger I production run.

Specifications for these standardized armor alloys were:

Armor Type	Spec. Date	Thickness	Resistance Kg/mm^2	Brinell Hardness	Alloy Percent
E32	Dec42	16-30 mm	105-120	309-353	3.1-4.2
E22	Feb43	35-50 mm	95-110	278-324	2.2-2.8
E22	Jun42	55-80 mm	90-105	265-309	2.2-2.8
"A"	Feb43	>70 mm	80-95	235-278	2.8-3.8
"B"	Feb44	All	75-90	220-266	3.5-4.7

Composition of each of the standardized armor alloys was a bit different, as shown in the following table:

	C	Mn	Si	Cr	Ni	Mo	V	P&S
E22	0.37-0.47	0.60-0.90	0.20-0.50	1.60-1.90	—	—	<0.15	<0.05
E32	0.37-0.47	0.60-0.90	0.20-0.50	1.20-1.60	1.30-1.70	—	<0.15	<0.05
"A"	0.32-0.42	0.60-0.90	0.20-0.50	2.00-2.60	—	0.20-0.30	<0.15	<0.05
"B"	0.32-0.42	0.60-0.90	0.20-0.50	2.00-2.60	0.70-1.20	—	<0.15	<0.05

Armor Steel Alloy Composition in Percentage

British examination of a captured Tiger I, **Fgst.Nr.250570**, revealed that by early 1943 none of the plates were face-hardened and all within the prescribed tolerance of -0% to +5% for thickness. These hardness readings may be slightly off, as they were taken with a Poldi portable hardness tester.

Plate	Thickness in mm	Approx. Brinell Hardness figure
Turret Roof	26	290
Gun Mantlet	100-200	280
Turret Side	82	255
Hull Roof	26	335
Driver's Front	102	265
Glacis Plate	62	265
Hull Nose	102	265
Superstructure Side	82	255-260
Hull Side	63	265
Hull Rear	82	255

PENETRATION TESTS

On 31 March 1942, Wa Pruef 6 reported on the results of penetration tests of the cast armor gun mantlet for the **1.Serie VK 4501 (P) (Tiger)** conducted at Kummersdorf on 17 March 1942:

The purpose of the trial was to test the gun mantlet produced by Fried.Krupp AG, Essen by directly firing at it with 5 cm and 7.5 cm ammunition.

The gun mantlet, made out of a special steel casting in accordance with drawing number 021 B 863-11 without installing parts, was mounted on a sled so that it was held at an angle of 0° to vertical.

1. Results of firing **5 cm Pz.Gr. Pak 38** *from a* **5 cm Pak K.u.T. (L/70)** *at a range of 100 meters.*

<u>1st Hit</u>: Underneath on the edge of the mantlet. Edge hit, not usable.

<u>2nd Hit</u>: About 70 mm above the first hit. No penetration. No other effect.

<u>3rd Hit</u>: Directly underneath the left hole bored for the gun sight. Clean penetration. Left hole squeezed shut.

2. Results of firing **7.5 cm K-Granate, rot. Pz., blind** *at charge 4 from an* **F.K.16 n.A.** *at a range of 100 meters.*

<u>4th Hit</u>: Directly under the third hit. Clean penetration.

<u>5th Hit</u>: Under the hole bored for the machinegun. Deflected penetration (nose of the shell visible). Corner broken off.

Conclusions: The gun mantlet can safely provide protection against hits from **5 cm Pzgr.38** *or* **Pzgr.39** *rounds fired at ranges of 100 meters or greater in all locations where it is not weakened by borings for the gun, gunsight, or machinegun. It does not provide safe protection against hits from* **7.5 cm Pz.-K.Granate, rot** *rounds fired at a range of 100 meters.*

Suggested Improvements: The weak points at the borings for the gun sight and the machinegun can be made secure by adding thicker bulges which connect to the center piece. Consideration must be taken of the field of view from the gun sight as well as heat conduction from the recoil enhancer of the machinegun.

The suggested modifications can't be incorporated by Krupp because all 100 gun mantlets of this type for the **VK 45.01** *turret have already been completed by Fried.Krupp A.G. for a contract from Dr.Porsche K.G. without previously involving Wa Pruef 6 in their design or method of production.*

For armor to be judged to have provided complete protection, it had to withstand a hit without resulting in a crack through which light could be seen, pieces being broken off the backside, or worse. Normally, 80 mm rolled plate was considered to be proof against hits from uncapped, inert **5 cm Pzgr.39** fired at an initial muzzle velocity of 850 m/s. However, it was proven by testing that 95 mm of cast armor "A" and 100 mm of cast armor "B" was needed to provide the same protection as 80 mm of rolled armor plate.

During testing of the **VK 45.01(P)** armor hull at Kummersdorf on 30 April 1942, Krupp representatives were embarrassed in front of Prof. Porsche, Dr. Rohland (D.H.H.V.), General von Rademeir, Oberst Fichtner (Wa Pruef 6), Oberstlt.v.Wilcke (Wa Pruef 6) and Baurat Rau (Wa Pruef 6) when the 102 mm thick driver's front plate at 8° was cleanly penetrated by a **7.5 cm Pzgr.** which resulted in numerous flakes and significant pieces broken out. Oberst Fichtner was deeply disillusioned. Baurat Rau declared that the armor was defective because other 100 mm thick plates at 0° had withstood hits from the same ammunition.

In a meeting on 7 May 1942 between Krupp representatives and Baurat Rau (head armor expert at Wa Pruef 6 IIb), the penetration of the driver's front plate in the **VK 45.01(P)** armor hull was discussed: *Penetration tests on a comparison plate of the same thickness and hardness made out of alloy* **PP794**, *fired at with* **7.5 cm K-Granate rot P** *(Vo 630 m/s, weight 6.8 kg) at 0° had resulted in three* **BmRoL** *(bulges with cracks without light). This comparison plate was also on the border of being defeated so that a guarantee can't be given that the driver's front plate won't fail to provide complete protection. Baurat Rau acknowledged this fact. Increasing the hardness is hardly possible even by changing the alloy. Also, thickening the driver's front plate for the first 100* **VK 45.01 (P)** *armor hulls can no longer be accomplished. Because in the interim the driver's front plate built into these hulls have been face-hardened, an increase in their resistance to penetration is expected. The PP793 plate, tested on 30 April 1942, had inclusions but these should not have appreciably reduced its resistance to penetration.*

Starting with **VK 45.01(P)** armor hull number 150050, some of the 80 and 100 mm armor plates were face-hardened, and from number 150060, all were face-hardened. On 2 June 1942 Krupp reported: *Penetration tests using* **7.5 cm Pzgr.** *were conducted on a* **VK 45.01(P)** *driver's front plate, held at an angle of 0.5°, made out of* **PP793** *alloy (0.43% C, 0.34% Mn, 0.30% Si, 2.26% Cr, 0.29% Mo, 0.02% P) face-hardened to 207 Kg/mm² (564 to 601 Rockwell C scale) with 109 Kg/mm² (321 Brinell) on the back. Results: 1* **BmRoL** *with little flaking of the face-hardened surface, estimated to be 7 to 8 mm deep and 1* **BmRoL** *in which a piece 20 mm by 40 mm flew off the back of the bulge. During the second hit the plate itself bent back about 40 mm. The tests have shown that face-hardening has improved resistance to penetration. However, even this did not result in complete security against all types of failure.*

Refer to Chapter 7 for the results of other penetration tests conducted by the British against Tiger I armor.

Appendices

APPENDIX E - COMPARISON TRIALS

On 14 October 1942, Krupp was informed: Assembly of the 100 **Porsche-Tigern** has been suspended until further notice. Comparative driving trials between both Tigers was to occur in Berka near Eisenach at the end of October, followed by a demonstration for the Panzerkommission on 2 November for the purpose of making a decision on continuing production, and a demonstration for Reichsminister Albert Speer on 3 November.

A further notice about the demonstration on 2 November 1942 revealed that in addition to the Tigers, other Panzers were to take part, including:

2 **Tiger (Henschel)** 1 normal, 1 with electrically shifted transmission
2 **Tiger (Porsche)**
1 **Panther (MAN)** with turret
1 **Panther (Daimler-Benz)**
1 **VK 3601** with adjustable driver's seat
1 **Pz.Kpfw.III ZW 40** with semi-automatic transmission

As recorded by a representative from M.A.N., the Panzer demonstration and driving trials took place in Berka from 8 through 14 November 1942:

*The following vehicles taking part in the demonstration were lined up in the following order: 2 **Henschel-Tiger**, 2 **Porsche-Tiger**, 2 **MAN-Panther**, 1 **Daimler-Panther**, 1 **Las 138** with electrically shifted transmission, 2 russian tanks (1 T 34, 1 KW 1), 1 ZW with wide tracks, 1 ZW with **Schachtellaufwerk**, 1 ZW with rubber-saving roadwheels, 2 ZW as **Flammenwerfer**, 2 **Panzerspaehwagen** outfitted with **Flammenwerfer**.*

After unloading, the vehicles were viewed by Oberstleutnant Holzhaeuer and other officers already at the train station. As we were told there, a further demonstration should take place in the afternoon. The vehicles had to be available on the grounds ready to start at 1400 hours.

Minister Speer arrived at 1530 hours escorted by General Fichtner, Oberst Thomale, Oberstlt. Holzhaeuer and other officers.

*All vehicles took part in the driving trials on this first day, and drove in part through difficult swampy terrain. The **VK 3601** bellied out in an especially difficult location and had to be pulled out.*

At 1000 hours on 9 November, Minister Speer and others from the Panzerausschuss continued the demonstration. The vehicles were driven in a row to Mittelberg. The climbing capability of several vehicles was tested in a deep muddy hollow.

*The **VK 3601** came first. In spite of renewed attempts, it didn't make it through the designated stretch but then drove through a less difficult location. Next, the **Henschel-Tiger** also couldn't drive through the designated stretch. Both of the **Porsche-Tiger**, next in line, drove through, but they had turned the guns to the rear.*

*The ground was very heavily plowed up before both of the **MAN-Panther** took their turn. In spite of this, both of the **MAN-Panther** drove cleanly through. The next vehicle was the **Daimler-Panther**. This vehicle didn't get through in spite of repeated attempts and was towed out. During this trial, the **Daimler-Panther** suffered damage to the transmission, fell out after driving about 2 kilometer across the meadow land, and didn't take part in any further driving trials.*

*Minister Speer requested that both of the **MAN-Panther** drive through again. But the ground was so heavily churned up that they couldn't drive through the designated stretch. **Panther 1** drove up the slope somewhat to the left and **Panther 2** drove up somewhat to the right. Also both of the **Posche-Tiger**, which were directed to try again, didn't get through.*

*After the hollow, the journey went through a very muddy forest trail to the meadowland. From here, **Panther Nr.2** didn't take part in further driving trials on this date. Instead, because of damage to the engine, it was driven back to the shop.*

*Minister Speer drove the **VK 3601**. All vehicles were lined up on the open meadow in order to test their speed. In driving a distance of 150 meters, which included climbing a small hill, the **VK 3601** led the **Panther** by about 10 meters. All other vehicles were about 50 meters behind. Because of damage to a cooling fan, the **Panther** couldn't be driven any further and was towed back by a **Porsche-Tiger**.*

*On 11 November, the vehicles were viewed by frontline officers. The vehicles were again driven to Mittelberg, where climbing trials again took place in the same hollow. On this day none of the vehicles made it through the hollow on their first attempt. The **VK 3601**, driven by Oberst Thomale, broke down during its try and couldn't be repaired. It didn't take part in any further driving trials.*

*Representatives from industry came on 12 November. The journey was again made to Mittelberg in order to again complete climbing trials in the same hollow. Both of the **Henschel-Tiger** didn't get through the designated location and had to drive through at a less steep location. Both of the **Porsche-Tiger** also didn't get through. One had to be pulled out. After the climbing attempts, the journey was made back to the starting point.*

*The vehicles were shown to frontline officers, this time troop engineers, on 13 November. The trip was again made to Mittelberg through the known hollow. A location was chosen that wasn't particularly difficult, so that all of the vehicles made it through. During the return trip, the vehicles weren't required to be driven in a row. The **MAN-Panther** easily passed all of the other vehicles and got back 10 minutes earlier than the rest. The following vehicles took part in this trip: 1 **Henschel-Tiger**, 1 **Porsche-Tiger**, 2 **MAN-Panther**, 1 **Las 138**, 2 **ZW**. All the rest of the vehicles had fallen out.*

APPENDIX F - CAMOUFLAGE PAINT AND ZIMMERIT

Camouflage paint schemes on the Tiger I went through the following four phases:

1. Originally, the Tiger I left Henschel completely covered with a base coat of **Dunkelgrau** (RAL 7021) paint. Tigers sent to Russia were to be covered with whitewash in the winter.

2. Tiger I, designated as "**Tropen**" (for service in Tunisia and with some units sent to southern Russia), were to be spray-painted in a two-tone camouflage scheme (two-thirds **Braun** (RAL 8020) and one-third **Grau** (RAL 7027)) at the Henschel assembly plant prior to being shipped to the ordnance depot for issue to the troops. Tigers issued to **Grossdeutschland**, **SS-LAH**, **SS-Das Reich**, and **SS-Totenkopf** in December 1942 through February 1943 were painted in the "**Tropen**" camouflage scheme.

3. Starting in February 1943, the Tiger I left Henschel completely covered with a base coat of **Dunkelgelb** (RAL 7028) paint, onto which the troops were to apply camouflage patterns using **Deckpasten** in **Olivgruen** (RAL 6003) and **Rotbraun** (RAL 8017) colors.

4. Starting in August 1943, **Zimmerit** anti-magnetic coating was applied and dried before the Tiger I was completely covered with a base coat of **Dunkelgelb** (RAL 7028) at the assembly plant. Troops were still to apply camouflage patterns using **Deckpasten** in **Olivgruen** (RAL 6003) and **Rotbraun** (RAL 8017) colors.

The **8.8 cm Kw.K. L/56** was to be coated with a lacquer (instead of paint) in the same base color as the rest of the Panzer

Units created their own camouflage patterns by spray painting **Olivgruen (RAL 6003)** and **Rotbraun (RAL 8017)** in irregular shapes onto the base coat of **Dunkelgelb (RAL 7028)** paint. (NA)

Authors Hilary L. Doyle and Thomas L. Jentz

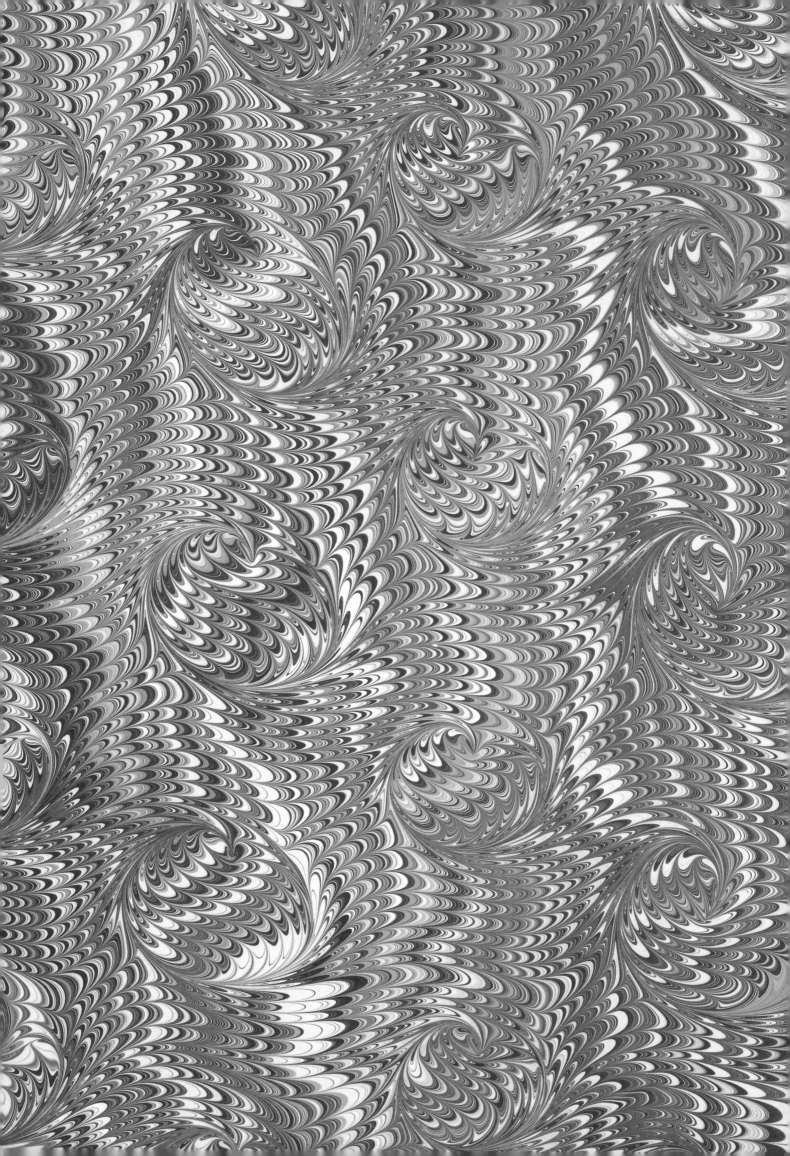